NATIONAL GEOGRAPHIC'S
HOW THINGS WORK

Interactive and virtually real, a video game works its 3-D magic.

National Geographic's
How Things

WORK

Wings and engines help a jetliner defy the force of gravity.

EVERYDAY
TECHNOLOGY
EXPLAINED

BY JOHN LANGONE

ART BY PETE SAMEK,
ANDY CHRISTIE,
AND BRYAN CHRISTIE

NATIONAL
GEOGRAPHIC
WASHINGTON, D.C.

CONTENTS

INTRODUCTION 6

AT HOME: RUNNING THE HOUSE

14

POWER AND ENERGY

52

BUILDING AND BUILDINGS

64

TRANSPORTATION: ON THE MOVE

80

SOWING AND GROWING

118

FABRICS AND FIBERS

130

142    AT PLAY: ENTERTAINMENT

178   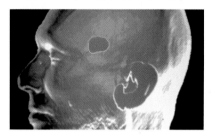 MANUFACTURING AND MINING

194    TOOLS OF MEDICINE

210    INFORMATION AND COMMUNICATION

246    OTHER WORLDS: NEAR AND FAR

AFTERWORD   258

GLOSSARY   262
ILLUSTRATIONS CREDITS 265
ADDITIONAL READING AND ACKNOWLEDGMENTS 266
INDEX   267
STAFF CREDITS AND AUTHOR'S NOTE 272

INTRODUCTION

IT has been said that whenever Thomas Edison showed visitors the numerous inventions and gadgets that filled his home, someone would invariably ask why tourists still had to enter through an old-fashioned turnstile. The master inventor had an easy answer: "Because every single soul who forces his way through that old stile pumps three gallons of water up from my well and into my water tank."

Versatility, as well as the ability to perform useful work, is what machines and mechanisms are often all about, and while a device may have a specific use, chances are that its operating principles and many of its parts, if not the device itself, can be put to work in a variety of ways. The compact disc, for example, fills our homes with stereophonic music, but it also can provide images, data, and sound that emerge from a personal computer. Propellers drive planes through the air, but they also help generate electricity in a hydroelectric power plant. Water irrigates soil-grown crops, and it can itself become the medium for growth on a soilless hydroponics farm. Fabrics let air in but also keep it out; lenses allow us to look deep inside our bodies and far into the universe. A home's furnace can also run a central air-conditioning system; an electron microscope works more like a television set than like a conventional microscope; the telephone expands into the Internet; the steam engine lends a hand to drive a nuclear-powered submarine; the levers that move a piano's keys can be modified to run a typewriter.

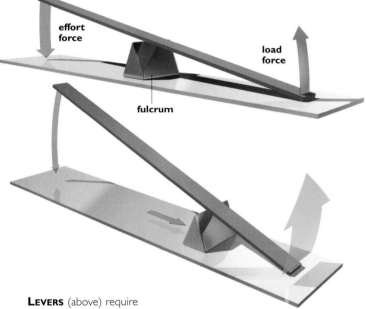

LEVERS (above) require effort at one end to raise a load on the other, and the effort required to lift the load depends on the position of the fulcrum. With equal amounts of effort and load at its ends, a lever balances when its fulcrum occupies a position at the center. Moving the fulcrum farther from the load (top) makes the load harder to raise. Placing it closer to the load (above) reduces the amount of effort needed for lifting. To help a **hammer** (right) overcome the resistance of wood, a wrist serves as the fulcrum.

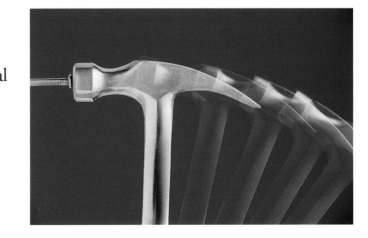

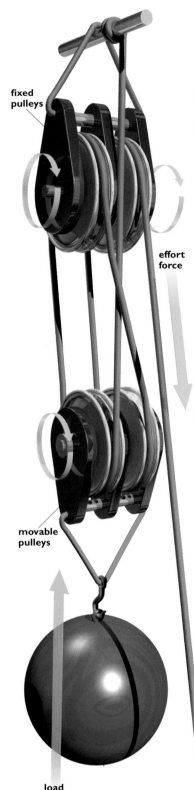

fixed pulleys

effort force

movable pulleys

load force

The machines and other technologies that have extended the range of human capability are all products of a human ingenuity that has successfully utilized the science behind motion, forces, and various forms of energy. One can look under the hood of an automobile or unscrew the back of a television set to check out the working parts, but to truly understand how a machine works—and a machine can be something as simple as a lever, an inclined plane, a wheel and axle, a pulley, or a screw—one has to have at least a casual acquaintance with nature's forces and how they are modified, transformed, transmitted, and otherwise adapted to do work.

In general terms, machines work because some kind of force, or energy that affects motion, is applied. Put another way, machines are devices that overcome resistance at one point by the application of a force at some other point. Various movable mechanical parts, such as levers, gears, and springs—or electrical wires, transistors, and electromagnets—manipulate and transmit the force exerted by a "prime mover," sending it where it is needed and getting the most out of it. This is where efficiency comes in. Always less than 100 percent because some energy is lost to friction as heat, a machine's efficiency is the ratio of useful work to the energy put into it, and it varies enormously depending on the machine. Only about 6 to 8 percent of the energy from burned coal went into the work the old steam locomotive did to haul the train. On the other hand, a gasoline engine and a steam turbine may have a thermal efficiency of around 25 percent, a diesel oil burner about 35 percent, a hoist some 60 percent, and an electric motor around 95 percent.

The process through which a machine works often involves interconnections: As a force moves one component of a machine, that part exerts force at another point, moving some

PULLEYS (left) may consist of grooved wheels mounted on blocks, with a rope running in the grooves. A fixed pulley changes the direction of force applied to a rope. If you run the end of that rope around an unfixed pulley and then pass it back to the fixed one, you can raise a load attached to the rope with half the effort. Various combinations of ropes and pulleys help distribute the weight of a **load** (above), easing the strain on workers who must lift heavy objects.

8

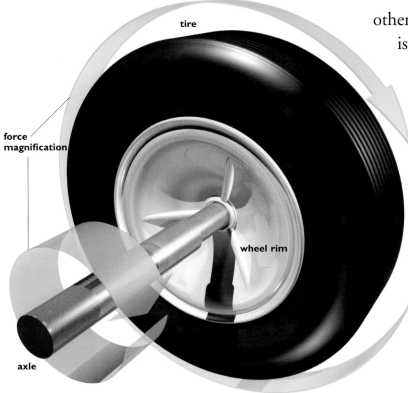

tire

force
magnification

wheel rim

axle

HOW A WHEEL WORKS
(above): With its familiar
circular frame, hub, and axle,
a wheel effectively transmits
power and motion, moving
everything from toys and
vehicles to houses and the
hands of a watch. Whether
spoked, smooth, toothed,
or flanged—or used in a
Ferris wheel (right) or a
wheelbarrow—the wheel
provides the rotary motion
essential to countless
machines. Its center acts
as a fulcrum, making it,
in fact, a rotating lever.

other part, and so on. Thus, energy
is handed along by the machine
in a sort of relay. In a nuclear
power plant, for example,
the force of heat from
split atoms vaporizes
water, creating the steam
that turns the turbine
that runs the electrical
generator. In an
automobile, vaporized
gasoline ignites and
burns inside the
cylinders, transmitting
motion to the wheels.
But the laws of force and
motion don't apply only to running
a steam engine or driving an automobile. In the human body,
chemical energy moves the muscles that do the work when we
swing axes or hit golf balls, while the bones act as levers, and
the joints are controlled by the body's equivalent of belts and
pulleys. Fired bullets and rockets follow the force-motion rules.
The water that rushes from a faucet, the furnace that forces heat
into a room, and the electrical pulses that actuate a loudspeaker
do so as well.

The simplest of our machines is probably the lever, a rigid
beam or rod pivoted at a fulcrum
point. Used in one basic form for
centuries, at first to lift heavy stones
or tree trunks, it is now used in
crowbars, wheelbarrows, nutcrackers,
typewriter bars, pliers, wrenches, nail
clippers, kitchen scales, and even the
links within a piano that move the
hammers that strike the tightened
wires to sound the notes. A pulley,

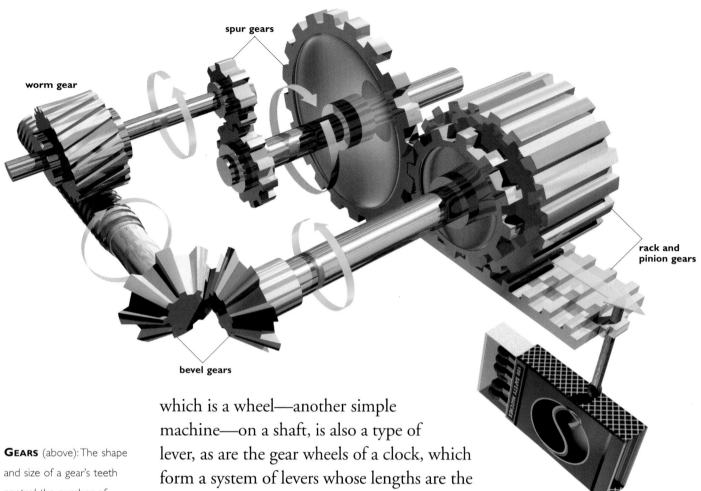

spur gears

worm gear

rack and
pinion gears

bevel gears

GEARS (above): The shape and size of a gear's teeth control the number of rotations, the direction of motion, the speed, and the amount of force exerted. Worm gears link shafts that have axes at right angles but which do not intersect; bevel gears connect shafts at an angle; rack and pinion gears have pinion wheels that mesh with sliding, toothed racks. Spur gears, essential to the workings of a basic **clock** (right), mesh with those on parallel shafts.

which is a wheel—another simple machine—on a shaft, is also a type of lever, as are the gear wheels of a clock, which form a system of levers whose lengths are the radii of the wheels. The inclined plane is a simple machine that enabled the ancient Egyptians to haul up stones to build the pyramids; its spiral cousins include the screw and the automobile jack. The operating principle of an inclined plane (effort to move an object to a higher plane is reduced by extending the distance the object must travel) also governs the action of the wedge, which translates work at one end to force at the sides, as in splitting wood.

Last, but certainly not least, are wheels and axles, which take advantage of the principle that rolling friction is less of a drag than sliding friction. Shoving a heavy box across a floor, for example, may be impossible without rollers of some kind beneath; with them, however, friction is dissipated and relatively little force is required. A power and motion transmitter par excellence, the wheel mediates between rotary and linear motion, stores energy as in a flywheel, and shows up in gears, winches, capstans, turbines, and pulleys.

The mechanisms and principles behind simple machines, and the complex ones that they create when mixed and matched, is what this book is all about. *How Things Work: Everyday Technology Explained* is not, however, a home repair book. It will not tell you how to fix a laser printer or tune a piano, for example. Nor is it a manual that will teach you how to pilot a plane or stitch a hem with a sewing machine. What it will do instead is satisfy your curiosity about how and why these machines work. It will also tell you about a host of other machines that affect our lives in various ways.

The pages ahead include more than information on conventional machinery. Other technologies—those that appear to be "nonmachines," for want of a better phrase—are equally interesting, and while they are not gadgets in the classic sense, they still have a place here because they, too, rely on scientific principles to function. Thus, the book not only explores what goes on inside a refrigerator, a computer, a jet engine, and most of our other familiar machines, but it also examines the technology and the principles behind such achievements as building arches, transgenic pigs, cloned sheep, nicotine patches, drug therapies, stress-reducing sports equipment, clarinets, fireworks displays, and soil-lifting detergents. Each one of these things works in a way that fulfills the definition of technology as any product of the applied sciences having

A MOTOR (right) converts electrical into mechanical energy—just the reverse of what a generator, or dynamo, does—and it works on the principle that an electrical conductor moves when in the presence of a magnetic field at right angles to the current. Current enters the motor via a commutator and its brushes, and it turns a laminated steel rotor that contains conducting coils. The commutator reverses the flow of current every half turn of the coils, and this action reverses the magnetic field, thus ensuring that the coil continues moving. As the rotor turns, a shaft delivers mechanical power impressive enough to move a **vacuum cleaner** (opposite) and any household pet in its way.

rotor

stator coils

commutator

brush

brush

shaft

practical value or industrial use. As Thomas Carlyle said in the 19th century, "It is the Age of Machinery, in every outward and inward sense of that word." His words are as true now as they were then.

How Things Work: Everyday Technology Explained addresses the most important mechanical, electrical, and electronic "things" in our lives, as well as their debt in varying degrees to magnetism, acoustics, chemistry, physics, biology, bioengineering, civil engineering, and agricultural science. The book is arranged, by theme, into 11 chapters. Each chapter includes a brief introduction followed by several lavishly illustrated two-page spreads focusing on various machines or technology. Cross-references on each spread will direct you to similar technologies, devices, processes, and principles described on other pages.

The book's range is eclectic. On one page you will see inside a baseball and a golf ball; another page will let you peer inside a gigantic magnetic bottle that contains the awesome spawn of fusion. A bar-code scanner and an ATM are viewed as most customers have never seen them; a computer, logic gates, and the Internet are stripped to their bare essentials; the hooks and loops of a Velcro fastener are unraveled; and special effects used in the film *Titanic* are explained. Household plumbing and electrical systems emerge skeleton-like when walls are removed. The mysterious mechanism beneath a clock's face and the powerhouse inside a nuclear submarine are opened to view. From an air bag to a climbing crane to kimono-patterning to a Zamboni, everyday technology

12

is not only explained but also rendered easy on the eyes. Within each chapter, sidebars give you insight into such topics as the pendulum, earthquake-protection, the motorcycle, bathroom fixtures, environmental cleanup, the zipper, high-tech haute couture, radios, and the electronic-music keyboard. They also help to highlight the application of science to specific technologies and will provide a grace note or two in the midst of some of the more technical discussions. You can open *How Things Work* at any place and read each single-topic spread as if it were a magazine article.

While the technology and scientific principles behind numerous devices and processes are generally explained, occasionally in detail, this book is not in any way meant as a textbook on engineering or physics. It attempts to make abstruse concepts understandable, but not so simple as to insult a reader's intelligence or to belittle the science. Moreover,

POWER STROKES (below and opposite, bottom left): In a four-stroke internal combustion engine, vaporized gasoline ignites and burns inside the cylinders. The downward induction stroke opens an inlet valve, sucking in fuel and air; on compression, an upward stroke compresses the air and fuel mixture; in the downward power stroke, a spark plug ignites the fuel; on exhaust, a valve opens, expelling gases. A diesel engine, like the one powering a **train** (opposite, top right), relies on heat produced by compression to ignite the fuel.

Induction

Compression

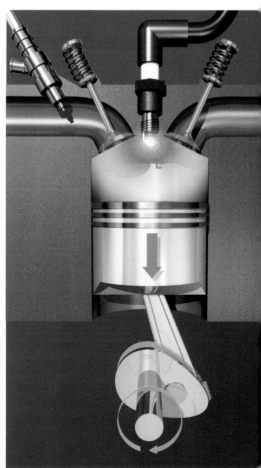

Ignition

because technology overlaps so many fields, the interdependence of the scientific, industrial, and technical disciplines involved in creating and utilizing our machines, equipment, and systems is always hovering implicitly or stated explicitly in the text. The overlap becomes especially obvious in the section describing the MRI scanner, which is a wedding of medicine, electronics, computer technology, and nuclear physics. You can also see it in the explanation of the kidney dialysis machine, which is a product of physics and bioengineering; and in the discussions of such industrial processes as the manufacture of steel, plastics, and glass, all of which involve the blending of basic science and technology.

Inventions and innovations are, of course, the creations of the human mind. Where possible, *How Things Work: Everyday Technology Explained* includes historical information about a gadget or a process, relating how, when, and why it came to be and describing its impact on society. We learn, for instance, that clocks may have been in use in 2000 B.C., that Leonardo da Vinci designed a steam-powered vehicle, that Queen Elizabeth I refused to grant a patent to the inventor of a knitting machine, and that miners were once paid by the tons of coal they loaded by hand.

In the last analysis, even a casual reading—and the book is designed for that purpose—should, it is hoped, increase one's awareness of the importance of science and technology in all of our lives. We might also come to appreciate an observation made by an 18th-century Polish king, who said that science, when well digested, is nothing but good sense and reason.

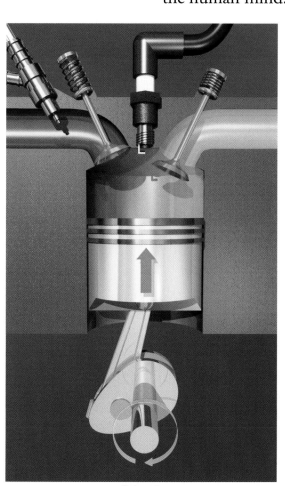

Exhaust

THE HOUSE

BE it ever so humble, there is no place like a modern home. It usually is outfitted with heating and cooling equipment, numerous appliances, and a variety of tools, and it is pierced by a circuitous system of wires, pipes, vents, and ducts. Today's comfortable, efficient structures rely on electricity and the transformer, a 19th-century invention that made it possible to transmit power for domestic as well as industrial use. Because we have electricity, houses now are filled with "home basics" people once considered luxuries. Such things include furnaces, fans, air conditioners, refrigerators, stoves, lightbulbs, washing machines, and various digital items; few of us would function well without them. The average home of today contains at least a hundred pieces of powered equipment—machines that can reduce our physical labors and help ease our minds.

Home owners can rest easy with today's array of high-tech appliances.

COOKING WITHOUT GUESSWORK

SEE ALSO

Home Heating • 24

Mediums and Messages • 214

Wired • 30

IN 1900, the Sears, Roebuck and Company catalog offered a top-of-the-line Acme sterling steel, nickel-plated range that burned coal or wood and sold for a then whopping $26.50 to $31.05. Today, a high-tech home cookstove with computerized precision controls can cost thousands of dollars, but a person doesn't have to pay anywhere near that amount to cook like Fannie Farmer or Julia Child.

Standard electric and gas ranges are efficient heat producers, while microwave ovens, toaster ovens, electric skillets, and slow cookers are good alternatives. Although each of the electrical cooking appliances has its own size, shape, and purpose, all but the microwave do their jobs by converting electricity into heat. As electricity passes through insulated wires inside metal coils or loops, called heating elements, the electrical resistance heats the outer metal. Manual controls regulate the heat by adjusting voltage in the wires, or a thermostat makes adjustments automatically. In a gas range, natural or bottled gas flows in and is mixed with air in a chamber; then the mixture is ignited by a spark, the flame of a pilot light, or an electric heating coil, firing the burners. A third variation is the convection oven, and in this appliance a fan circulates heated air uniformly and continuously around food for faster and more even cooking.

A microwave oven produces high-frequency electromagnetic waves. Passing through food, the waves reverse polarity billions of times a second. The food's water molecules also have polarity, and they react to each change by rapidly reversing themselves. Friction results, heating the water and cooking the food.

Suspended animation (left): Heat, latches, timers, and springs pop toasted bread out of a toaster.

French fries in bubbling oil (below left) turn crisp and brown under the direction of a hidden heating element and a watchful thermostat that maintains an even temperature.

Cooking with gas (below): Easily controlled gas chars a thick steak in a cutaway view. Less efficient than electricity, a gas burner loses much of its heat to the surrounding air.

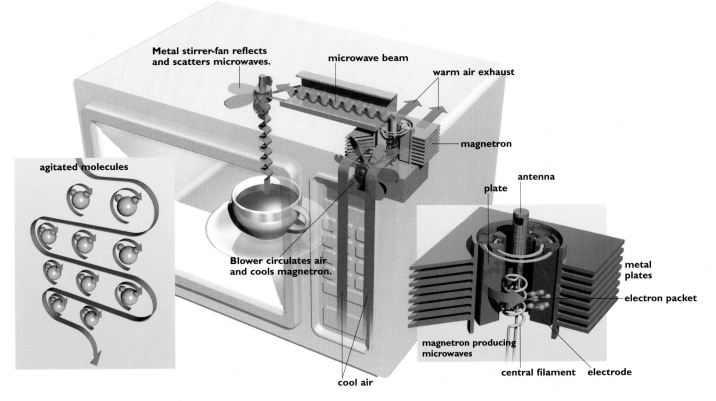

Metal stirrer-fan reflects and scatters microwaves.

microwave beam

warm air exhaust

magnetron

agitated molecules

plate

antenna

Blower circulates air and cools magnetron.

metal plates

electron packet

magnetron producing microwaves

cool air

central filament   electrode

A MICROWAVE OVEN (above) uses the same electromagnetic radiation as radar to magnetically agitate **WATER MOLECULES IN FOOD** (left inset), causing them to heat faster than in a conventional oven. Produced electrically in a **MAGNETRON** (above and right inset), the high-frequency microwaves (shown in purple) pass through a wave-guide, encounter a stirrer-fan (green), and reflect into the oven. Invisible and seemingly benign, microwaves can injure human tissue.

Range of possibilities

(right): A standard and relatively inexpensive appliance in many home kitchens, a gas-fired range mixes air with natural or bottled gas ignited by a spark or a pilot light.

THE FRIDGE

SEE ALSO

**Fans and Air
Conditioners
• 28**

Refrigerated food (above right): Chilled air keeps food from deteriorating, a principle recognized long ago when people stored foodstuffs in cool caves. Cold air slows the growth of bacteria but, contrary to popular belief, does not kill them.

The first Frigidaire (right) came off a 1921 assembly line at the Delco Light Plant, a subsidiary of General Motors. Eventually, Frigidaire products found a place in one out of four U.S. homes.

IN the days of the early icebox, the simple process of ice melting inside the enclosure took up a certain amount of heat and kept the icebox cool. If someone covered the block of ice with paper to conserve the coolant, the box would not function properly because the paper kept the ice from melting.

The modern refrigerator also cools by extracting heat, but it does so in a more complex way. A compressor pumps coolant vapor through sealed tubes; after increasing its pressure and temperature, it routes the vapor outside the refrigerator box and into a condenser where the coolant releases heat and becomes a liquid. From there, pipes lead the liquid back into the box, through a control valve, and into the evaporator, or freezing unit; it vaporizes in the coils surrounding the unit and absorbs heat, thereby cooling the unit. The warmed vapor returns to the compressor, and the cycle begins again. In a frost-free refrigerator, a fan in the freezer compartment circulates air cooled by the evaporator coils, helping to prevent condensation on the freezer walls. Two or three times a day, a timer activates a defrost heater that melts frost on evaporator coils. Water drains into a pan at the bottom of the refrigerator and slowly evaporates into the room.

Of course, the reason for all of this is to preserve food. Although the cold air does not kill the bacteria and molds that are present, it checks their growth and slows the chemical breakdown of food. A freezer compartment is typically 0°F to 10°F, cold enough to preserve food up to a year, and the refrigerator enclosure itself is usually between 32°F and 40°F.

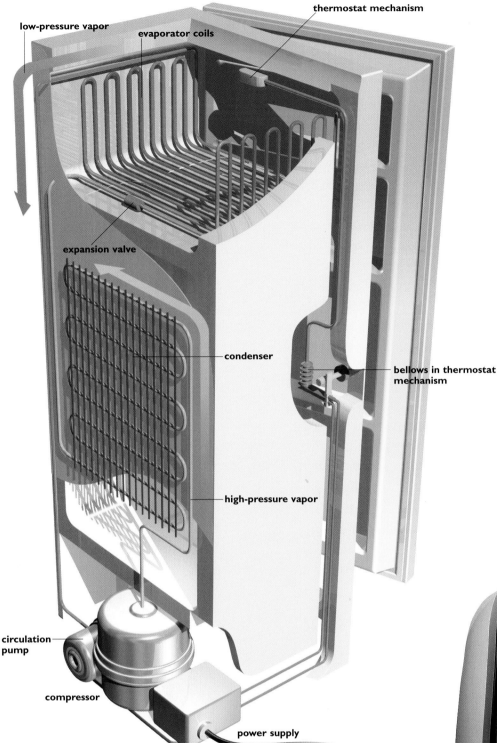

low-pressure vapor

evaporator coils

thermostat mechanism

expansion valve

condenser

bellows in thermostat mechanism

high-pressure vapor

circulation pump

compressor

power supply

COLD STORAGE (above): Like an old-fashioned icebox, a refrigerator also keeps food cold. But this common modern appliance does not require a big block of ice melting inside to help it accomplish its task. Instead, a refrigerator relies on coolant circulated through a sealed system of tubes by a compressor. As the coolant evaporates, it absorbs heat from food stored inside the refrigerator compartments. The warmed gas returns to the compressor, which sends it to the condenser where it cools and becomes a liquid. Then the coolant recirculates through the sealed system.

Fridge wizardry (below): Conceived by Italian designer Roberto Pezzetta, the four-and-a-half-foot-tall Oz refrigerator has a door that rolls on a rubber ball and a chlorofluorocarbon-free coolant. Its creators call it "a bivalve shell, offering the food it contains without reserve, without secrets."

SMALL APPLIANCES

SEE ALSO

**Cutting
Edges**
• **46**

**Workshop
Tools** • **44**

SMALL household appliances generally require fairly strong motors that are capable of varying speeds. Most such appliances rely on universal motors, which are versatile power units that operate on either direct or alternating current. To fulfill its role, an appliance also has a unique combination of gears, cams, drive belts, pivots, shafts, rotors, and pinions. These connect the motor itself, directly or indirectly, to blades that shave, slice, and dice, and to other attachments for beating, cooling, blending, and keeping time.

This combination of parts makes even the humblest kitchen appliance an ingenious mechanism that demonstrates and uses complicated principles of physics, mechanics, and engineering. Appliances often contain gears, which transmit motion and power from one part of a machine to another and may be called upon when more twisting and turning force than the motor ordinarily provides is needed. Reduction gears slow an appliance when too much speed is undesirable. Motors may also be used to drive cams or other mechanisms that translate rotary motion into linear motion.

Many small appliances do not need gearing because their attachments work best when they are drawing full power directly from the motor. For example, while an electric mixer cannot be a direct extension of the motor's shaft because the speed would make a mess of the mix (a mixer employs a worm gear to slow up the action), a fan or the cutting blade in a food processor works directly off the motor, spinning at the same high speed.

direction of
rotation

blade screen

head

circle of
blades

driveshaft

ROUND AND ROUND:

Many household appliances rely on revolving wheels and gears. Clamping the handles of a **MANUAL CAN OPENER** (below) forces a cutter to bite into a can. A toothed wheel grips the rim of the can and drives the can around as the turnkey rotates. Gears on the turnkey mesh with gears on the cutting and toothed wheels, imparting rotational force. An **ELECTRIC SHAVER** (above right) rotates a circle of blades on a springlike driveshaft that follows the contours of the skin. A **GARBAGE DISPOSAL UNIT** (far right) grinds refuse on a turntable before flushing it through a drainpipe.

Double duty (opposite): Beauty and a meal go hand in hand as a resourceful woman dries her hair, 1960s style, while giving a kitchen dish a nonchalant stir.

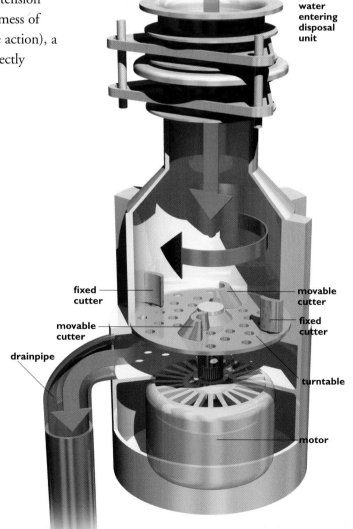

handle lever

handle lever

circular cutter

can

turnkey

toothed wheels
(or gears)

waste
and
water
entering
disposal
unit

fixed
cutter

movable
cutter

movable
cutter

fixed
cutter

drainpipe

turntable

motor

PIPES, PUMPS, AND PLUMBING

SEE ALSO

Construction Elements • 66

Home Heating • 24

Solar Heating • 26

Washing and Drying • 34

HISTORIAN Will Durant tells us that Romans carried plumbing "to an excellence unmatched before the twentieth century." Pipes made of lead brought water from aqueducts and mains into homes; fittings and stopcocks were of ornamented bronze; and leaders and gutters carried rainwater from rooftops.

A modern domestic plumbing system, more of a maze than anything the Romans ever constructed, is essentially an inner skeleton of main and secondary pipelines broken in places by shutoff and bypass valves, faucets, traps, toilets, showerheads, tanks, clean-out plugs, water heaters, and drains. Receiving and discharging water are, of course, the reasons for plumbing. Water from a municipal supply or private well usually enters a home under pressure and is routed through hot- and cold-water lines to the fixtures. When the water must be hot—for the bathroom shower or the automatic dishwashing machine, for example—it enters from the cold-water main and flows through a cold-water inlet into a gas-fired or electric water heater. In a thermostat-controlled tank, it is heated to the proper temperature and then sent to the hot-water faucet via a separate outlet pipe. Cold-water taps, along with the toilet, are linked directly to the cold-water main.

Unlike the intake system, the drainage system is not pressurized; it relies on gravity to dispose of water that was flushed from a toilet or let out of a sink. Fixtures are connected to a large drainpipe by curved traps filled with water; the water acts as a seal to keep potentially dangerous sewer gases from flowing into the home. At the bottom of the system, a main drain transports waste to a town sewer line or to a septic tank on the property.

In and out: Each day, we rely on plumbing fixtures to bring in clean water and to take out wastes. The handy **faucet** (top) admits water under pressure from pipes linked to a private well or to a municipal water supply. Water in a **toilet bowl** (above) also enters through a pressurized intake system, and it drains out, when flushed, through pipes that carry it to a nearby septic tank or to a town's sewer line.

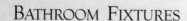

BATHROOM FIXTURES

Classic clutter aside, a modern bathroom is a testimonial to the science of water conveyance. From simple sinks to elaborate Jacuzzis, from fancy showerheads to toilets with warmed seats and electronically operated flushes, fixtures for the room known as the "john," the "gents," the "ladies," and the "head" are a plumber's (and a designer's) dream. The basic flush toilet brought comfort, convenience, and sanitation to dwellings by transferring the outhouse to our house. An extraordinary combination of inflow and outflow, it releases water from a tank into a bowl, from which it whirls down and out through a water-filled trap to a drain.

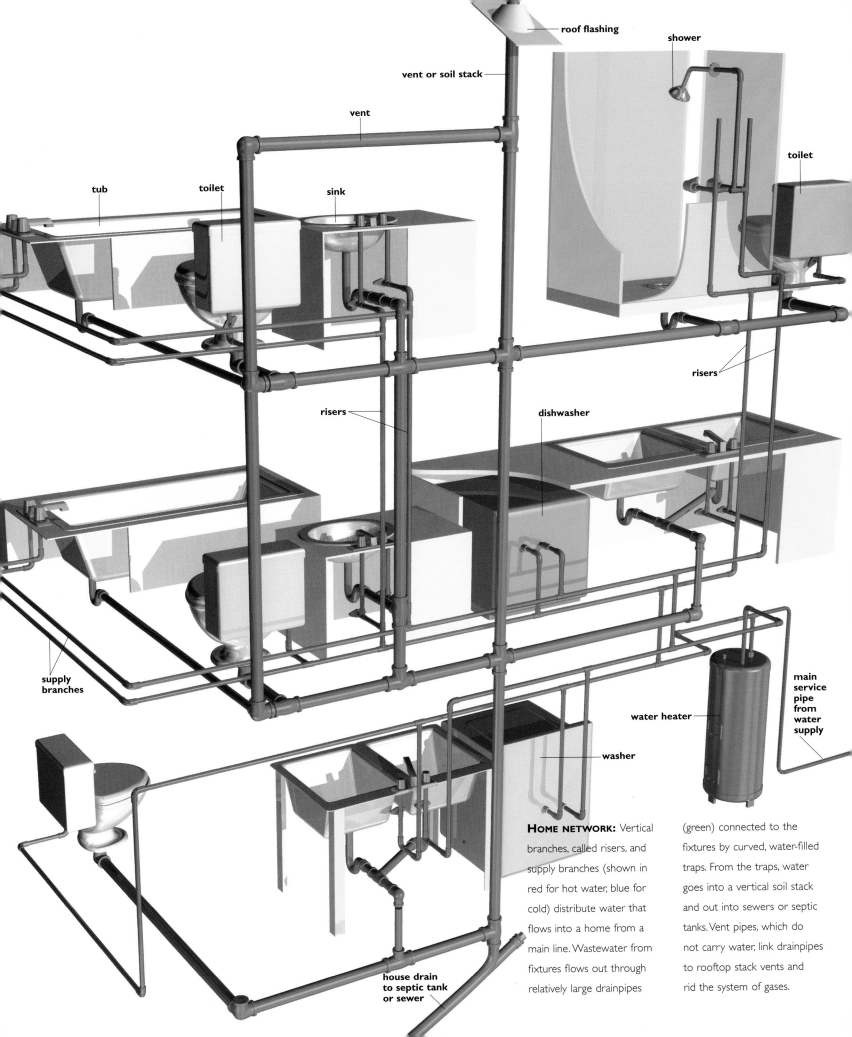

roof flashing

shower

vent or soil stack

vent

toilet

tub

toilet

sink

risers

risers

dishwasher

supply
branches

main
service
pipe
from
water
supply

water heater

washer

house drain
to septic tank
or sewer

HOME NETWORK: Vertical branches, called risers, and supply branches (shown in red for hot water, blue for cold) distribute water that flows into a home from a main line. Wastewater from fixtures flows out through relatively large drainpipes (green) connected to the fixtures by curved, water-filled traps. From the traps, water goes into a vertical soil stack and out into sewers or septic tanks. Vent pipes, which do not carry water, link drainpipes to rooftop stack vents and rid the system of gases.

HOME HEATING

SEE ALSO

Fans and Air Conditioners • 28

Solar Heating • 26

Wind and Geothermal Power • 62

TO warm their homes, the Romans devised a heating system that used wood- or charcoal-burning furnaces to supply hot air to rooms through tile pipes or channels under floors and inside walls. Centuries later, the occupants of medieval castles had no such system. For all their imposing appearance and the romantic images they evoke, the castles were dark, dreary, uncomfortable places. Not until the late 1700s did Benjamin Thompson, a British physicist, introduce heating improvements that led to the modern world's first central heating system.

Gas and oil burners are at the heart of today's home heating systems, which work by heating air or water and circulating the heat throughout the house. In the 1950s, oil burners began replacing coal-burning furnaces. Although they are more expensive to install than gas burners, the oil they use is often cheaper than gas. Natural gas burns more cleanly than oil, but this fuel is not available in many places and therefore not always an option.

Whether oil or gas is the fuel of choice, both must be burned, and both need to be mixed with air to do so. In a gas furnace, the gas-and-air mixture is ignited, and the heat is used to produce hot water, steam, or hot air. A forced-air or hot-water circulating system spreads the heat throughout the home as waste gases go up a flue. An oil burner also sets fire to a mixture of fuel and air, and in this type of system the fuel is first pumped from a tank and converted into a fine spray. It is then burned in a combustion chamber.

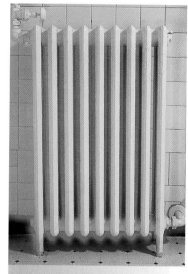

RADIATORS

Radiators do just what their name implies: They radiate. The one shown above radiates heat from the steam generated by a basement boiler. Once standard fixtures in many homes, the heavy metal radiators hissing heat into a room have given way to convector units—devices containing water pipes, fins, and grilles—which disperse the heat from circulating hot water.

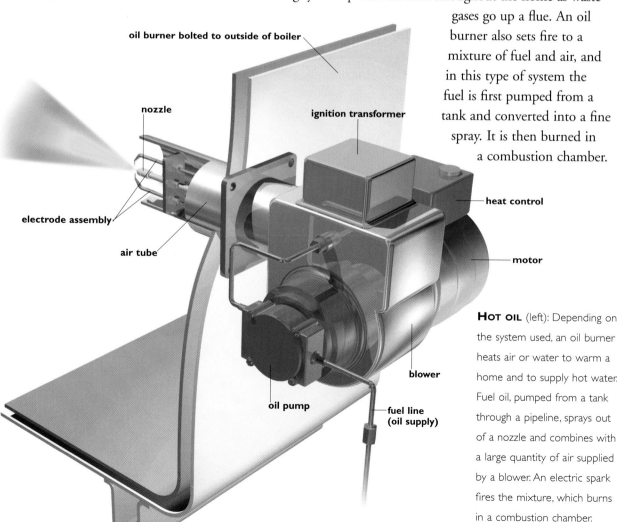

oil burner bolted to outside of boiler

nozzle

ignition transformer

electrode assembly

air tube

heat control

motor

oil pump

blower

fuel line (oil supply)

HOT OIL (left): Depending on the system used, an oil burner heats air or water to warm a home and to supply hot water. Fuel oil, pumped from a tank through a pipeline, sprays out of a nozzle and combines with a large quantity of air supplied by a blower. An electric spark fires the mixture, which burns in a combustion chamber.

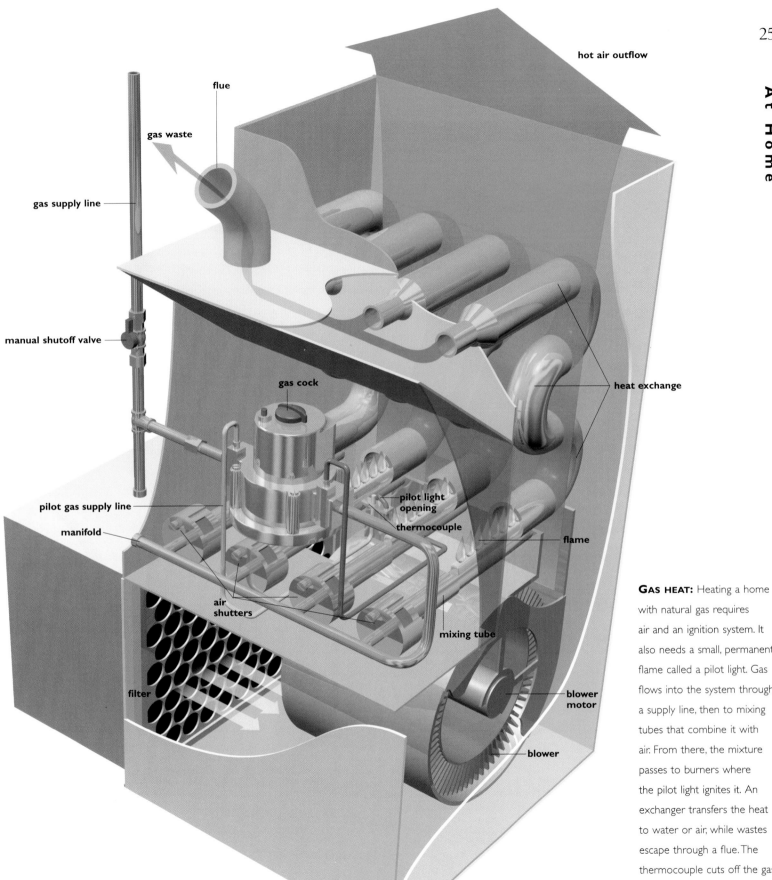

hot air outflow

flue

gas waste

gas supply line

manual shutoff valve

gas cock

heat exchange

pilot gas supply line

pilot light opening

thermocouple

manifold

flame

air shutters

mixing tube

filter

blower motor

blower

GAS HEAT: Heating a home with natural gas requires air and an ignition system. It also needs a small, permanent flame called a pilot light. Gas flows into the system through a supply line, then to mixing tubes that combine it with air. From there, the mixture passes to burners where the pilot light ignites it. An exchanger transfers the heat to water or air, while wastes escape through a flue. The thermocouple cuts off the gas flow if the pilot light goes out.

SOLAR HEATING

SEE ALSO

Alternatives
• 86

Home Heating
• 24

Pipes and Pumps • 22

FOSSIL fuel supplies will not last forever, a fact that underlines the importance of conserving resources and finding alternative energy sources to heat our homes. One option is solar power, the energy emitted by the sun as electromagnetic radiation. It is plentiful, nonpolluting, and free.

Today, water or air heated by the sun fulfills a variety of household needs, providing hot water for kitchen sinks and showers or warm air to take the chill out of the house itself. Capturing solar energy to heat a home can be done in two ways: passively or actively. Passive heating depends on a structure's design, materials, and location to maximize the warming effect of sunlight. The method works best in a well-insulated home with large, south-facing windows (in the Northern Hemisphere), an interior of dark stone or tile, and a location in a sunny region. Such a home receives lots of sunlight, which warms the air inside; the dark interior absorbs sunlight and reradiates the heat during the night.

The active method requires equipment that collects, stores, and distributes solar energy. A flat-plate collector fixed to the roof of a house serves either an air- or a water-based system. A cover made from a clear material, usually glass, allows sunlight to pass through and strike a dark, heat-absorbing metal plate. Warmed air between the cover and the plate helps to insulate the collector. As the plate soaks up heat, it warms air or water contained in pipes or tubes inside the collector. If the system is air-based, a fan blows the warmed air into a large, rock-filled storage area under the house. A water-based system pumps the warm water into an insulated storage tank for household use or through tubes placed under floors and in ceilings to provide space heating.

Sun worship (opposite): Resembling windows in a cathedral, the solar panels on this house collect the power of the sun on bright, clear days. Greater use of nonpolluting solar power could reduce the energy drain caused by buildings, which account for a third of all U.S. energy consumption.

Solar water heaters (below) perched atop a roof in Turkey provide an adequate supply of hot water for a household's needs. Pumps, pipes, and heat exchangers move the water from the collectors.

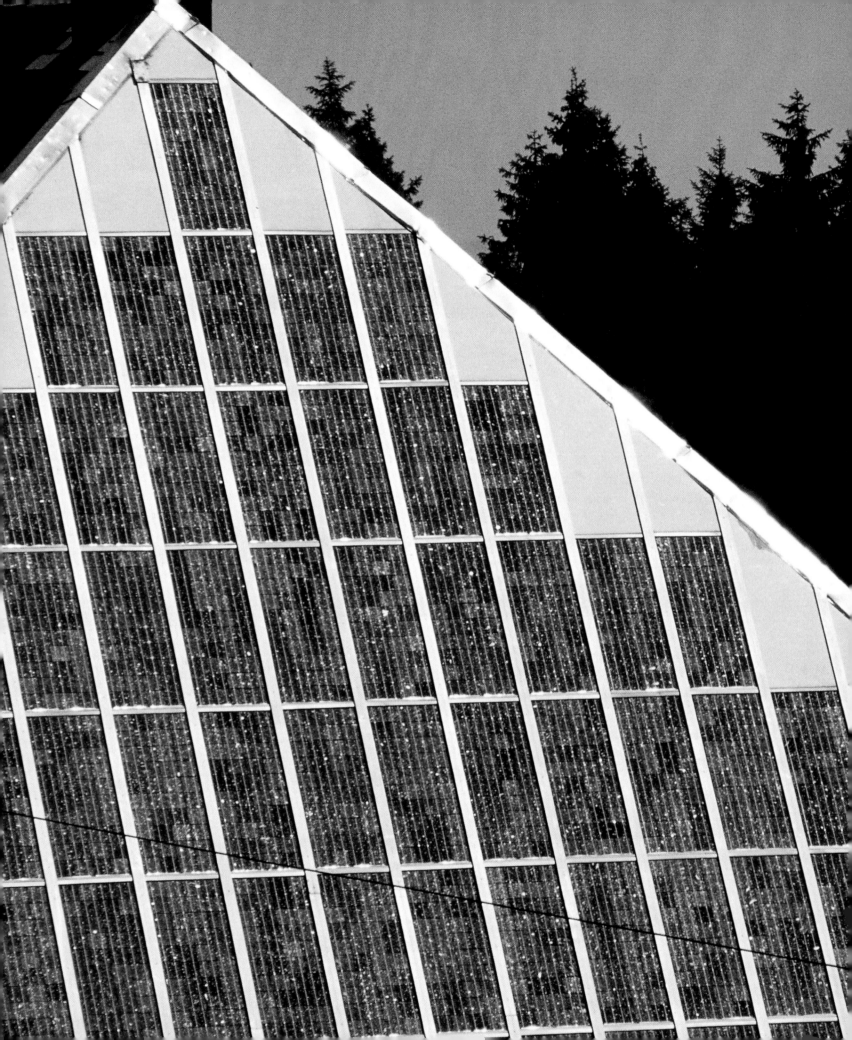

FANS AND AIR CONDITIONERS

SEE ALSO

Fridge • 18

**Home Heating
• 24**

**Solar Heating
• 26**

IN the days of silent movies, electric fans blew air over cakes of ice to keep theaters cool in summer. Fans themselves do not cool, however. They move warm or cool air around, which feels better than still and sticky air, or they push hot air out of a warm room or attic. But when fans are used with refrigerants, the comfort zone expands because the coolants dissipate unwanted heat.

Window units and central air conditioners remove heat and humidity from enclosed spaces and eject it outdoors, reducing the air temperature inside with the aid of a closed system containing a refrigerant. All such systems are based on these principles: When a liquid vaporizes, it draws heat from its surroundings; when a gas condenses back to a liquid under high pressure, heat is released.

With only a few modifications, central air-conditioning can work through a home's forced-air heating system. In this "split system," the evaporator coil is mounted on the furnace, and the condensing unit rests on a concrete slab just outside the home. A small tube carries liquid coolant in from the condensing unit, and a larger tube returns the coolant in a gaseous state to the unit outside. When the cooling system is in operation, the furnace blower draws in warm air from the house and sends it over the evaporator. As coolant in the evaporator changes from a liquid to a gas, it absorbs heat, thereby cooling the air. This process also causes the air to lose moisture, which flows into a condensation tray and out through a drain. The blower forces the cooled, dehumidified air into the supply ducts, and from there it circulates through the house.

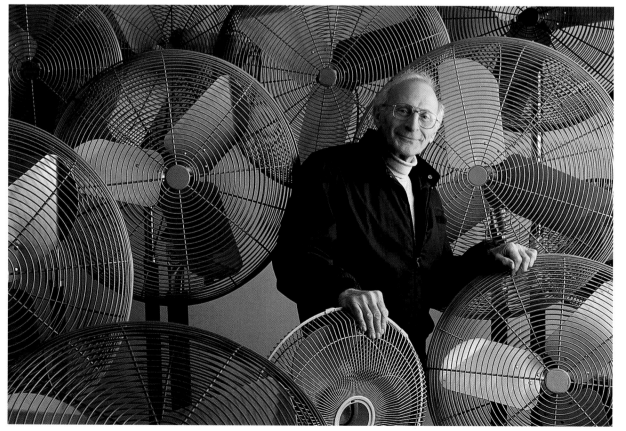

Blast from the past (above): Though the electric fan doesn't have the cooling power of an air conditioner, it does give the impression of coolness. The air current it produces feels more comfortable than still air.

Floor models (left) and their kin—tabletops, ceiling fans, and window units among them—meet many household needs. Fans in an attic or on a roof force hot air out and draw fresh air in; in bathrooms, laundries, and kitchens, they provide necessary ventilation.

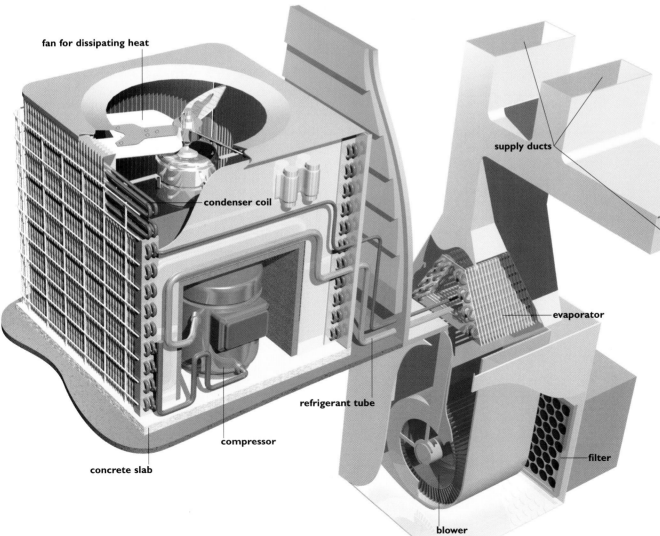

fan for dissipating heat

condenser coil

supply ducts

evaporator

refrigerant tube

compressor

concrete slab

blower

filter

Window air conditioners, from early units (left) to more recent models, work in essentially the same way. All mimic refrigerators in how they remove heat from a closed space and expel it.

CENTRAL AIR-CONDITIONING (left) dispenses comfort on command. A concrete slab outside the house holds the condensing unit, which pumps cooled, liquid refrigerant through tubes connected to the evaporator in the house. Inside, a blower sends warm air from the home over the evaporator. As the coolant changes from liquid to a gas, it absorbs heat and cools the air. A tube returns the coolant to the unit outside, while the blower forces the cooled air through supply ducts and into the home.

WIRED!

SEE ALSO

Making Electricity
• 54

Smart Buildings
• 72

IF a home could be stripped of its walls and framework, leaving the electrical system intact and upright, the remains would resemble a tangled and surreal sculpture of cables, wires, switch boxes, outlets, and dangling light fixtures. Invisible, yet flashing through it all, would be charged particles called electricity, which Benjamin Franklin referred to as a kind of fluid existing in all matter and whose myriad effects could be explained by too much or too little of it.

Probably our most versatile form of energy, electricity—with its whizzing electrons—is the lifeblood of a home, powering appliances and systems that are plugged into wall outlets or run on batteries. When supplied by a power company, electric current enters a home through overhead or underground cables. It goes through a meter that measures how much electricity is used and then flows to a service panel that distributes it over a wire network strung throughout the house.

Order is maintained by the service panel and the circuits—the continuous path that electricity follows through copper and aluminum conductors and other components. While the main power lines usually enter a house at one place, the power they carry is split and shunted out of the service panel over branch circuits, which are monitored by fuses and circuit breakers that are designed to prevent overloading, fire, and shock. The service panel generally contains three wires. Two of them are 120-volt wires that allow a household to run 120-volt and 240-volt appliances. (Voltage is actually the electromotive force with which a source of electricity sends electrons along a circuit to form a current.) The third wire is a zero-volt neutral wire; it connects to a ground wire attached to a metal pipe that goes into the ground. The cables forming branch circuits may hold two 120-volt wires, a neutral wire to complete the circuit, and a safety ground wire to protect the system when overvoltage occurs.

OVERSEER OF CURRENTS (below): The main service panel in a home wiring system controls the flow of electricity. Its switches serve as circuit breakers, automatically flicking off when a circuit overloads. Cables and wires lead to junction boxes, protective enclosures that house receptacles; grounded, the boxes send dangerous accumulated charges directly into the Earth. Outside the panel, switches turn lights on, off, or down, or they delay breaking a circuit, letting people leave a room before it darkens.

receptacles

junction box

switch

main service panel

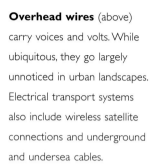

Overhead wires (above) carry voices and volts. While ubiquitous, they go largely unnoticed in urban landscapes. Electrical transport systems also include wireless satellite connections and underground and undersea cables.

Inside job (above): Visible for only a while and essential to our daily comfort, safety, and convenience, much of the work performed by an electrician remains hidden behind walls, floors, ceilings, and picture frames.

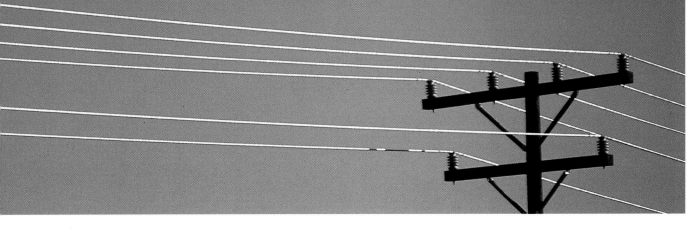

WIRING SYSTEM ANATOMY:
Overhead or underground cables bring electric current into a building. The service panel distributes it throughout the house over a network of circuits running inside ceilings and walls. Protected by circuit breakers or fuses, the circuits carry 120 or 240 volts regulated by a pair of 120-volt hot wires that transport power from the source; a third, or neutral, wire returns current to the panel, completing the circuit. A meter connected to the panel measures the flow of electricity in kilowatt-hours (kwh). One kwh represents a thousand watts consumed in an hour.

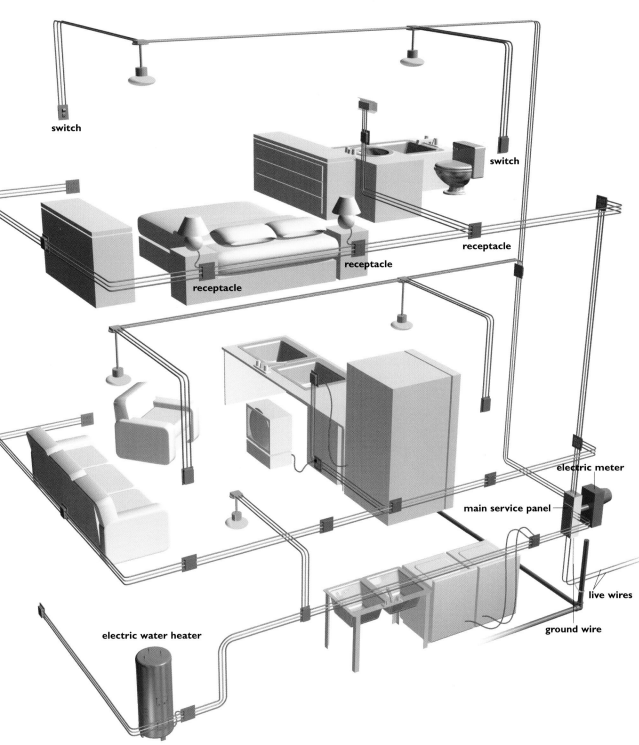

switch

switch

receptacle

receptacle

receptacle

electric meter

main service panel

live wires

ground wire

electric water heater

Let There Be Light!

See also

Making Electricity • 54

Mediums and Messages • 214

Wired • 30

Before Thomas Edison's incandescent electric lamp gave the world a new form of luminosity, people relied on brushwood torches, candles, oil lamps, and the light that erupted across the sky during a storm. Today, we brighten our homes with lightbulbs and fluorescent lamps. We find our way in the dark with flashlights, and we use streetlights, traffic lights, and automobile headlights to get around our neighborhoods. In our homes, new lightbulbs filter out unwanted colors and enhance vibrant ones in decorations and furnishings. We can raise or lower the lights to create just the right mood for a party or an intimate supper. Lights also help houseplants grow, readjust our internal body clocks after long international flights, and soothe seasonal affective disorders.

Lighting has evolved in astonishing ways since Thomas Edison invented his lamp. In 1879, he passed electricity through a strand of carbonized cotton sewing thread, causing the rudimentary filament to glow for more than 13 hours in a glass vacuum tube. His feat was marred, though, by the loss of power through heat and by the short bulb life.

The coiled tungsten filament, a metal with a high melting point that easily accommodates the heat needed for the best light, made a brighter, longer-lasting bulb. The life of a bulb got even longer when an unreactive gas—nitrogen, argon, or krypton—was placed inside to slow the filament's evaporation. A variation on the theme is the halogen lamp. Inside its bulb are molecules of bromine or iodine, halogen elements that combine with tungsten given off by the filament to form a gas. When this gas comes in contact with the hot filament, the tungsten atoms separate from the halogen and adhere to the filament, essentially rebuilding it.

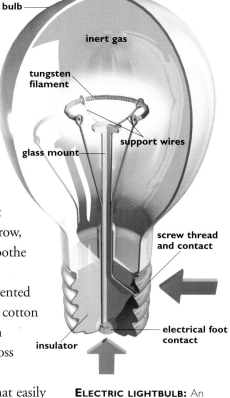

bulb

inert gas

tungsten filament

glass mount

support wires

screw thread and contact

electrical foot contact

insulator

Electric lightbulb: An incandescent bulb (above) works on the principle that a body—in this case, a thin tungsten wire filament—gives off visible light when heated to high temperature. Current passing through the filament heats the wire to more than 4000°F, resulting in the radiation of electromagnetic energy. An inert gas that fills the bulb reduces the possibility of the filament's oxidizing.

At the center of a halogen lamp (left) sits a hot filament. Tungsten evaporates from the filament, unites with a halogen compound inside the bulb, and forms a gas. When the gas touches the filament, it breaks down and releases tungsten atoms that rebuild the filament. More tungsten evaporates, and the cycle starts again.

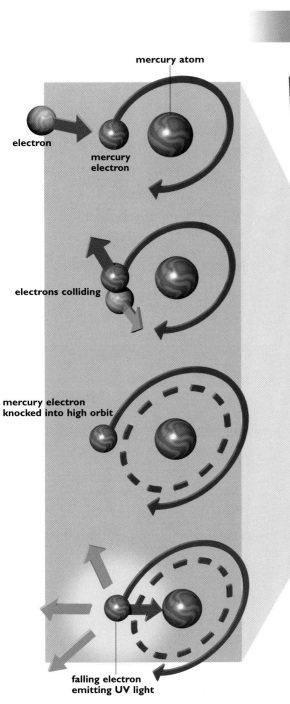

electric contacts

mercury atom

electron

mercury electron

electrons colliding

mercury electron knocked into high orbit

electrode

electrode

falling electron emitting UV light

electric contacts

BOTTLED LIGHT (above):

The light of a fluorescent lamp comes from the interaction of electrons and mercury vapor atoms. Pins pass current to an electrode, which emits electrons that strike atoms of vapor. This bombardment produces ultraviolet radiation that bounces off the phosphor coating, a fluorescent powder on the inside of the tube, energizing electrons in the phosphor's atoms. These atoms, in turn, radiate white light. Although far more complex than ordinary filament lamps, fluorescent lamps last longer, perform more efficiently, and produce much more light per watt of power consumed.

NEON

Lady Liberty (above) glows with the light of rare gases. As electricity streaks through her tubular construction, its flow takes on various hues during encounters with neon, which glows orange-red if sparked. Other gases cause other colors: Argon glows blue and red; krypton, bluish-white; and xenon, blue. Tinted tubes or tubes with coatings of luminescent powder enhance the effect.

SEE ALSO

**Fans and Air
Conditioners**
• 28

Home Heating
• 24

**Pipes and
Pumps** • 22

Cotton threads (above) come clean in an electron micrograph. Greatly magnified, these woven threads of a shirt collar appear free of grease and dirt after sloshing about in a washing machine filled with water and detergent. If washed by hand, the threads would not look as clean.

Wash day (right): Better than a washboard but no match for a modern machine, a wringer-type washer had a hose that emptied wastewater into the kitchen sink. Feeding clothes through rollers in the wringer squeezed out excess water.

BEFORE the versatile washing machine was invented, people used washboards to scrub clothes, or they carried their laundry to riverbanks and streams where they beat and rubbed it against rocks. Such backbreaking labor is still commonplace in parts of the world, but for most home owners the work is now done by a machine that automatically regulates water temperature, fills, measures out the detergent and bleach, washes, rinses, spin-dries, and empties. With its complex electrical, mechanical, and plumbing system, the washing machine is one of the most technologically advanced examples of a large household appliance. And when one considers the new models that operate with far less water, detergent, and energy, this machine seems to be even more of a wonder.

Manufacturers produce two types of washing machines: top-loaders and front-loaders. The top-loading models, which are generally more popular, rely on motor-driven agitator units that wrap and twist the clothes around during wash and rinse cycles. Pumps circulate, recirculate, and discharge the water, while the timers, switches, and sensors regulate the flow and temperature of the water, as well as the spin process. Top-loaders generally use an average of 40 gallons of water for each load of wash.

Front-loading machines are built without agitators. Instead of dragging and shoving clothes through a cycle, these machines churn the laundry about in high-speed, rotating drums. According to their manufacturers, front-loaders use 20 to 25 gallons of water for each washload.

The washer's companion is the dryer—electric or gas powered— and, like electric and gas furnaces, this machine circulates the heat from coils or burners to do its job. A fan draws air into the dryer where it is heated, passed through spin-damp clothes, and forced through a lint trap. The air is then ejected through an exhaust vent.

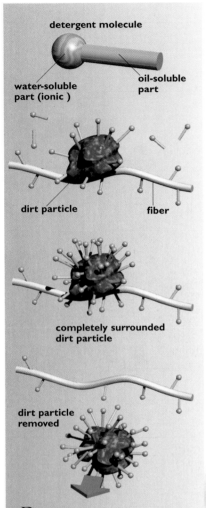

detergent molecule

water-soluble part (ionic) **oil-soluble part**

dirt particle **fiber**

completely surrounded dirt particle

dirt particle removed

DETERGENT

A detergent molecule has one "nonpolar" and one "polar" end. The nonpolar end (green) attaches to oily dirt while the polar end (blue) dissolves in the surrounding water, ionizing to form positive charges. These charges repel the ones on other soap-encrusted dirt particles, keeping the dirt suspended and allowing it to be rinsed away.

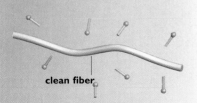

clean fiber

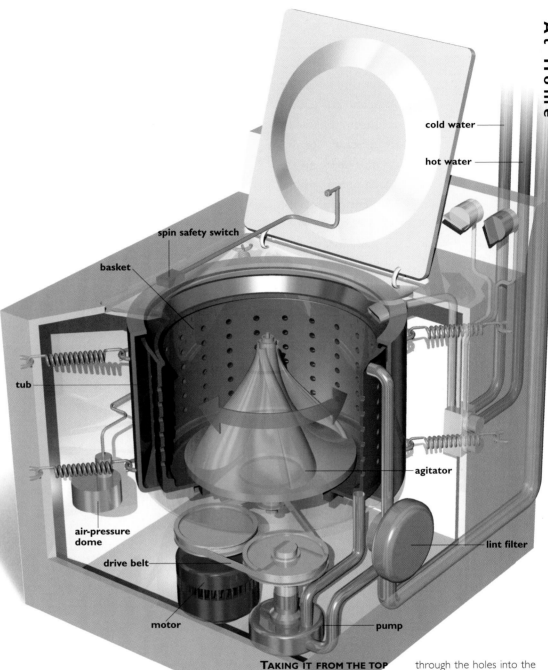

cold water

hot water

spin safety switch

basket

tub

air-pressure dome

drive belt

motor

pump

agitator

lint filter

TAKING IT FROM THE TOP (above): The cone-shaped, motor-driven agitator in a top-loading washer moves clothes back and forth in a tub filled with soapy water. When the wash cycle ends, a timer directs the motor to free the agitator and spin a perforated inner basket. Centrifugal force spins water through the holes into the outer tub and presses the clothes against the sides of the basket. After the water drains out, valves open and refill the tub for the rinse cycle. A final spin damp-dries the laundry. Springs attached to the tub and the unit's frame keep the washer stable during the unavoidable vibration.

DUST-BUSTERS

SEE ALSO

Fans and Air Conditioners
• 28

Small Appliances
• 20

Wright Stuff
• 110

IT'S IN THE BAG (below): In the scientific sense, a perfect vacuum means a complete absence of matter within a space—a rare, if not impossible, phenomenon. In a more familiar sense, a perfect vacuum describes a powerful cleaning machine that removes a great deal of matter from our rugs. An upright unit uses a partial vacuum that fills a bag with so much dirt that it requires frequent emptying. Air blown through the unit by a motor-driven fan escapes through the bag's porous walls, but dirt remains trapped inside.

THE carpet sweeper used to be (and still is in some homes) a good way to clean rugs and carpets. A simple device, it consists of an enclosed pan on wheels pushed about at the end of a long stick; crumbs and lint are scooped into the pan by the roller action of a cylindrical brush mounted on the underside. Although the sweeper is not much better than an ordinary dustpan and brush, it does not require stooping and kneeling.

Electric vacuum cleaners, which made their debut early in the 20th century, use suction, not gathering, to pick up particles sweepers just run over. Three basic systems—upright, canister, and central—all work in essentially the same way: A motor-driven fan creates a partial vacuum that sucks up dirt loosened by a beater brush and deposits it in a bag or another type of collector. (Vacuum cleaner fan-blades spin as much as 18,000 times a minute; jet engine blades reach 7,000 to 8,000 a minute.) An upright vacuum is a single unit with a bag usually attached along the back side of the handle. The canister type has a long, detachable cleaning wand attached to a rolling unit that houses a more powerful motor, the fan, and the dust bag.

A central vacuum cleaner has a power unit and a collection canister in a basement, garage, or spare room that is connected to metal or plastic tubes in walls and under floors. To activate the system, a person plugs a cleaning hose and its attachment-tool into inlet valves in the walls. Dust and debris are sucked through the tubes and sent to the dirt-collection canister, which has to be emptied only two or three times a year. More powerful and much easier to maneuver than a portable vacuum cleaner, a central unit keeps noise at a distance, and it doesn't spray out any of the fine dust particles that flow through a bag's porous walls.

dust bag

motor

beater brush

fan

drive belt

Vacuuming (right), viewed stroboscopically, seems almost a snap. Sold with numerous attachments, modern vacuum cleaners harness suction power to make housecleaning easier. They dust, remove dirt from crevices, clean upholstery, and even do walls.

YOUR BASIC CLOCK

SEE ALSO

**Electronic
Clocks • 40**

**Small
Appliances
• 20**

HOW did people tell time before they had clocks? They observed stars at night, the shadow of a sundial's gnomon, marks made on a candle, water that trickled through a clepsydra (water clock), or the grains of sand that sifted through an hourglass. Although we can still do it in those ways, it helps to have a plain old watch to make the necessary adjustments and put everything in sync.

Numerous people have claimed credit for inventing the true clock—the Chinese around 2000 B.C. and 11th-century Germans, for example. The oldest surviving mechanical clocks, however, date from late 14th-century France. While the principles behind early clocks still apply to modern timepieces, such innovations as atomic clocks, illuminated dials, jeweled bearings, quartz clocks, and digital alarms with snooze settings make the early versions seem closer to ancient water clocks than to our hex bezels.

Frills and variations aside, clocks are essentially boxes containing several toothed wheels, pins, and springs, all precisely arranged to measure that impalpable continuum called time. Because time is an ongoing, regular process, measuring it requires physical mechanisms that can keep pace with it. In the basic clock, power is supplied by electrical impulses or by a falling weight or an unwinding spring that can be rewound when necessary. Movement of the hands is regulated by a train of toothed wheels, pinions, and spindles of different sizes, each taking its own appointed time to make a full revolution. In some clocks, a pendulum controls the wheel train's rate of rotation. In other timepieces, a mainspring turns a driving wheel that moves the minute and hour wheels. These wheels are connected to hands on the face of the watch or clock. The minute wheel turns so that the minute hand takes one hour to go once around the face. (In 12 hours, it turns 12 times.) The hour wheel turns so that the hour hand takes 12 hours to complete one revolution.

Formidable timepiece,

a classic alarm clock (above) stands ready to rouse a sleeper with a nerve-shattering jangle. A powerful mainspring runs the alarm mechanism and the rest of the clock, and an alarm "hammer" does the waking by swinging back and forth against metal gongs.

PENDULUM

A pendulum's swing—the to-and-fro motion of a suspended weight—has a steady, measurable time span. Because successive swings occur in equal lengths of time, regardless of whether the swing is large or small, pendulums are ideally suited for controlling clock movements. Increasing or decreasing the length of a pendulum extends or shortens the duration of its swing, an adjustment that makes the period of the swing correspond to a desired time interval. Linked to a clock's escape wheel (see diagram on opposite page), which is gear-connected to the hands, a pendulum lets the wheel move one tooth for each swing.

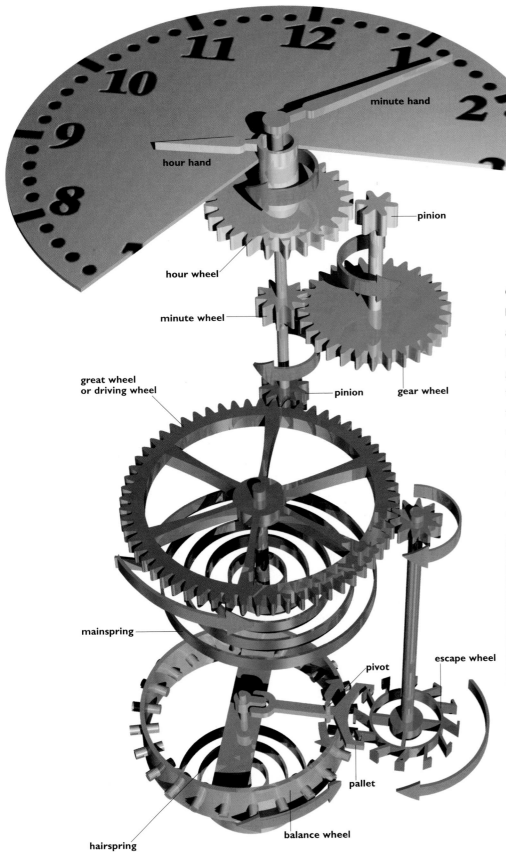

minute hand

hour hand

clock face

pinion

hour wheel

minute wheel

pinion

gear wheel

great wheel or driving wheel

mainspring

pivot

escape wheel

pallet

hairspring

balance wheel

CLOCKWORK (left): Powered by an unwinding mainspring, a clock also depends on the linkage of precisely machined gears and on a hairspring that winds and unwinds to control the movement of the balance wheel, which regulates the escape wheel. A main driving wheel meshes with the teeth of a gear wheel and turns as the mainspring unwinds, controlling the movements of the minute and hour wheels. The number of their teeth determine the number of revolutions that the hands will make.

Toothy spur gears (below) transmit the force of the mainspring and control the relative amount and speed of the hand movements.

ELECTRONIC CLOCKS

SEE ALSO

Basic Clock
• 38

Chips and Transistors
• 238

WHEN a small electrical charge is applied to a quartz crystal, the crystal begins to vibrate and give off pulses of current in a precise, predictable manner. These pulses, in turn, can be used to control the motor turning a clock's hands or to advance the numerals displayed by the liquid crystals in a digital display.

Such quartz vibrations are known as a piezoelectric reaction, and when they are put to work in a clock or a watch, they make for a timepiece that is far more accurate than a mechanical one. Couple the pulses of current from the oscillating quartz with a microchip that reduces their frequency to a usable rate—one pulse a second—and you have a clock or a watch that relies on miniscule electronic circuitry instead of toothed wheels and a mainspring.

While a quartz-crystal clock has an error rate of less than one-thousandth of a second per day, the latest generation of atomic clocks can be accurate to plus or minus one second in ten million years. Atomic clocks are faceless and handless wonders that trade on the constancy of the frequency of a molecular or atomic process. Put another way, atomic clocks use the extremely fast vibrations of molecules or atomic nuclei to take the measure of time. Because the vibrations remain constant, these clocks measure short intervals of time with much more precision than a mechanical clock.

GOOD VIBRATIONS (below): To keep time, a quartz clock relies on a common mineral and a microchip. Quartz, the mineral, can produce precise pulses of current when excited by a battery, and the microchip regulates the action. Electricity from the battery makes the quartz crystal vibrate at high frequency, while the chip reduces the rate and selects an appropriate frequency. The power controls the motor and gears that turn the hands of the clock; in a digital display, it advances the numerals that tell the time. The capacitor stores the electric charge. Some wrist models replace the battery with a tiny generator activated by the wearer's movements.

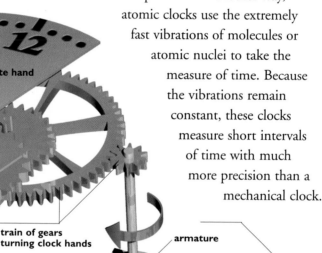

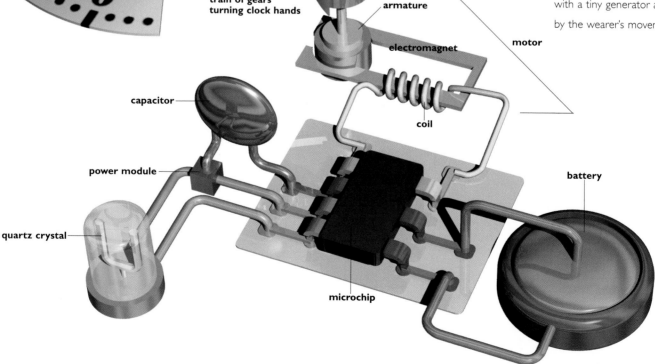

minute hand

hour hand

train of gears
turning clock hands

armature

electromagnet

motor

coil

capacitor

power module

battery

quartz crystal

microchip

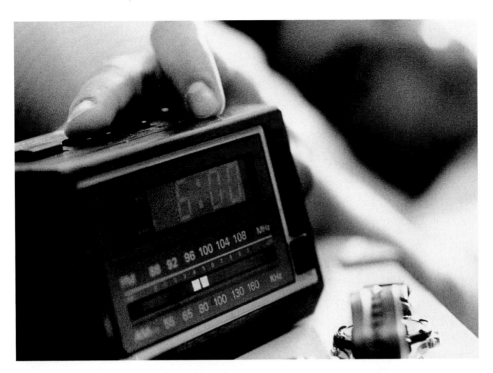

Alternating current powers a digital clock (upper left), and its precise cycles mark time's passage. Seven light-emitting diodes make up each digit and form numbers when energized in the correct combination.

An atomic clock (left) runs on a cylinder of cesium, a highly reactive metal whose atoms vibrate with exceptional constancy. Scientists measure these vibrations to come up with a primary time reference that has incredible precision. The International Committee on Weights and Measures, for example, now defines a second as the time an electron takes to spin on its axis inside a cesium atom; this translates into 9,192,631,770 oscillations.

SENSORS AND DETECTORS

SEE ALSO

Camera • 160

Nuclear Power • 58

Smart Buildings • 72

Wired • 30

OUR natural gifts of sight, smell, hearing, touch, and taste are not necessarily enough when it comes to accurately measuring some of the things that transpire around us. So, scientists have come up with a spate of devices that help us extend what is innate and ensure the safety and good health of ourselves and our families. Such devices include barometers, thermometers, pressure gauges, scales, Geiger counters, water and gas meters, seismographs, spirometers, and sphygmomanometers (for measuring blood pressure).

Smoke alarms and security systems exemplify the world of artificial senses. Smoke alarms "smell" fumes in different ways, mostly by interacting on a molecular level with the particles that make up the smoke. An ionization alarm is armed with a bit of radioactive material, and this material gives off atomic particles that form ions, which are charged atoms or molecules. The ions carry current in the detector, and when they interact with smoke particles, the result is a drop in the flow of current and the setting off of the alarm. A photoelectric detector works by beaming a steady light against a dark surface inside; when smoke seeps in, it scatters the light to a photocell that sets off the alarm.

Security systems use photocells to measure light levels so that they can turn lights on and off. They also detect infrared radiation given off by the body of an intruder. To detect an object's density and motion, other devices rely on reflected sound waves that are above the range of human hearing.

ON GUARD: One type of burglar alarm (opposite, lower left) uses a magnet on a door or a window for its trigger. If the door or window remains closed, the magnet continues to attract a metal bar fixed to the adjacent wall, maintaining an electrical circuit. Raising the window or opening the door, however, breaks the contact and sets off the alarm.

PIN-TUMBLER CYLINDER LOCK (left): By blocking the shear point between a lock's plug and shell, the drivers and pins keep a plug locked in place. If the right key enters the lock, the pins settle into its notches, clearing the shear point and letting the key rotate the plug.

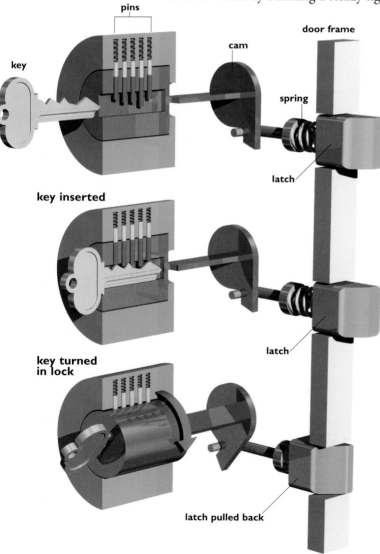
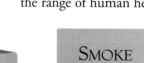

pins

key

cam

door frame

spring

latch

key inserted

latch

key turned in lock

latch

latch pulled back

SMOKE DETECTOR

Two kinds of smoke detectors sense smoke and sound alarms to protect people from fire. A photoelectric detector (above) contains a chamber in which a light source and a sensor are at right angles to each other, rather than directly opposite. If smoke enters the chamber, it scatters particles of light, causing some of them to hit the sensor. In an ionization detector, a tiny amount of radioactive material is used to ionize the air in a chamber and create a current. Smoke particles attach themselves to the ions and disrupt the current, setting off the alarm.

Pushing buttons (right):
To activate a security alarm system, a home owner usually has to do little more than program the system's sensor, telling it to set distance limits or to detect solid objects or their movements.

Masked but not hidden,
a stealthy intruder (below) would get stopped in his tracks by modern technology. Many of the sophisticated security systems that home owners rely on today come equipped with sensors that detect infrared radiation given off by the human body.

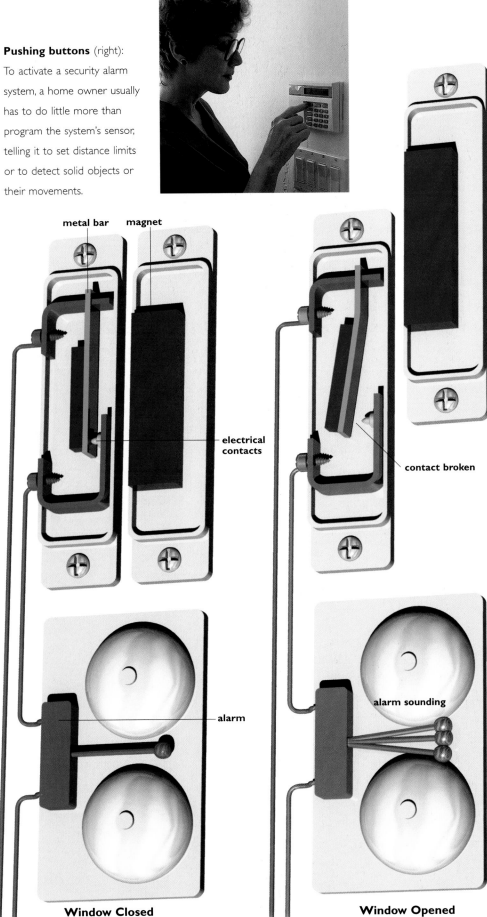

metal bar   magnet

electrical contacts

contact broken

alarm

alarm sounding

Window Closed

Window Opened

SEE ALSO

**Cutting Edges
• 46**

**Making
Electricity • 54**

**Small
Appliances • 20**

Wired • 30

Old reliables (right): Even basic hand tools get upgraded. Tough carbon steel now forms the business end of hammers and wrenches that produce little heat when used; steel tape measures have locks that control retraction; levels have unbreakable plastic vials. Some glues cure by drying, while others, such as cyanoacrylate superglue, cure by chemical reaction.

WHEN a home owner turns or twists a drill or a screw, he or she applies force to what is essentially a lever and, in so doing, produces torque, a rotational force. The effectiveness of the torque-producing effort depends on both the force exerted and the length of the lever's "arm." Thus, the handle of a screwdriver becomes more than a grip: It magnifies the force the hand uses to turn the blade and drive in a screw. The same thing is true of a pair of pliers twisting a nut. Both amplify the manual strength applied.

Electrically driven power tools, such as drills and screwdrivers, draw extra strength from a nicely meshed system of gears attached to a chuck, a jawlike fixture that holds the drill bits and the screwdriver blades. The tools also enjoy variable speed because of regulators that control the flow of electricity through the brushes to the motor's rotating armature.

Like cordless toothbrushes and shavers, cordless shop tools depend on small, efficient motors and batteries, those ever present devices that convert stored chemical energy to direct-current electricity. Nickel-cadmium batteries that deliver about 1.3 volts have been partly responsible for the popularity of cordless appliances. Also, such tools can be easily recharged by snapping them into stands linked to household current. The current—its voltage reduced by a small transformer in each unit—is applied in the direction opposite the one in which it flows when power is dispensed, a restoration process taking several hours. Some battery-driven appliances can be plugged directly into a wall outlet, a feature that allows them to receive constant charging when not in use.

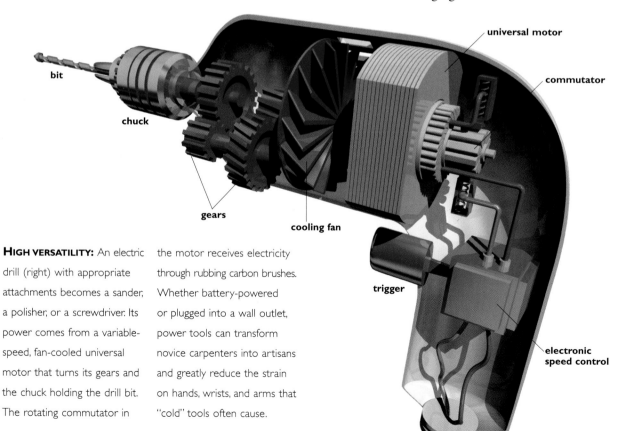

bit

chuck

gears

cooling fan

trigger

universal motor

commutator

electronic speed control

HIGH VERSATILITY: An electric drill (right) with appropriate attachments becomes a sander, a polisher, or a screwdriver. Its power comes from a variable-speed, fan-cooled universal motor that turns its gears and the chuck holding the drill bit. The rotating commutator in the motor receives electricity through rubbing carbon brushes. Whether battery-powered or plugged into a wall outlet, power tools can transform novice carpenters into artisans and greatly reduce the strain on hands, wrists, and arms that "cold" tools often cause.

CUTTING EDGES: SAWS AND LATHES

SEE ALSO

**Small
Appliances
• 20**

**Trade Tools
• 68**

**Workshop
Tools • 44**

ROTARY motion—the turning of an object around a center point, or axis—may be more important to machines than parts that move in linear fashion. Without it there would be no wheels, no gears, no belts, no motors, and no enormously useful tools such as electric saws and lathes.

A handsaw, as everyone knows, is a simple cutting instrument whose blade has sharp, pointed teeth that abrade wood or other materials. Saber saws and other electric versions mimic the back-and-forth motion of a handsaw, while requiring circular gears and rotating motors to do so. In some saws, the blade itself is circular and rotated by an electric motor. These power saws, which are very useful in the construction industry, have a variety of teeth, from coarse to fine, and can quickly saw large quantities of wood. Chain saws are portable saws with blade-studded steel links that form an endless chain. Powered by a gasoline engine or an electric motor, a chain saw runs its cutting chain on an oval guide bar at a high rate of speed.

A circular saw uses a sharp, round wheel to cut a straight line, but a lathe uses rotating motion to turn wood, metal, and other materials into cylindrical, tapered, and conical objects. The material to be shaped is held at each end of the machine and rotated against a cutting tool.

The thin blade of a jigsaw (opposite, top right) moves up and down to cut curves and to crosscut, bevel, and begin a cut in the middle of a wood panel.

Popular and portable, a circular saw (opposite, top left) has a motor that rotates the blade at high speed. The heavy-duty blade adjusts for depth and angle of a cut. A blade guard springs into safety position when the saw finishes its work.

right-hand cutter

drive link

Chips fly as a carpenter shapes wood fixed to a lathe (left). This basic turning tool works by rotating an object, such as wood or metal, about a horizontal axis. A cutting tool then moves across or parallel to the rotational direction, shaping the object as it turns.

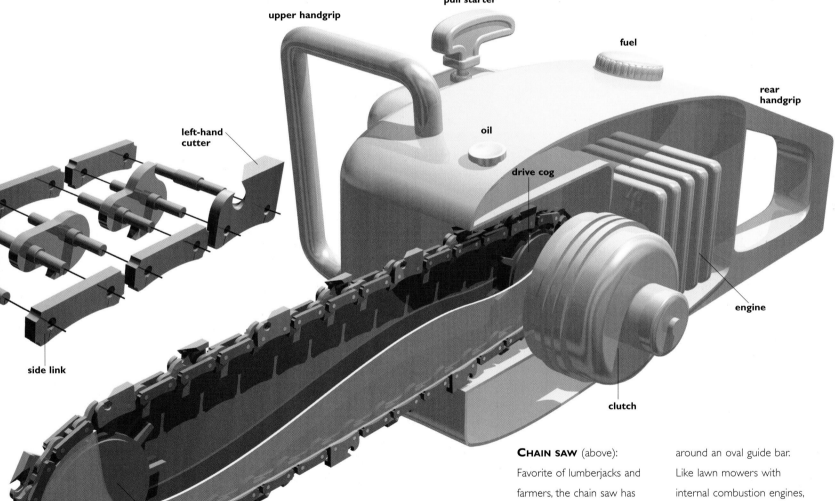

upper handgrip

pull starter

fuel

rear handgrip

left-hand cutter

oil

drive cog

engine

side link

clutch

tip sprocket

guide bar

chain carrying cutting teeth

CHAIN SAW (above): Favorite of lumberjacks and farmers, the chain saw has no match when it comes to mechanical brawn and mobile cutting power. Driven by a gasoline engine or an electric motor, the chain saw relies on its dominant feature: the blade-studded, bicycle-like chain (see close-up at left) that hurtles around an oval guide bar. Like lawn mowers with internal combustion engines, the gasoline versions require fuel and oil tanks, spark plugs, mufflers, carburetors, and clutches. Electric models run on universal motors; shafts turn gears that drive sprocket-and-clutch arrangements to move their chains.

GRASS CUTTERS

SEE ALSO

Alternatives
• 86

Cutting Edges
• 46

Workshop Tools • 44

BECAUSE a well-manicured lawn is a status symbol in many parts of the world, people will often invest a great deal of effort in controlling the growth of grass around their homes. Once, hand-swung scythes and grazing farm animals did the job, but nowadays—unless a home owner relishes the exercise provided by an engineless reel-type mower—a gasoline- or electric-powered machine spins the single blade that does the cutting and even the mulching.

A mower's blade runs parallel to the ground. Its raised rear edge sets up a draft that lifts blades of grass for cutting and blows the clippings out of a chute. In a mulcher-mower, grass clippings and leaves remain pressed against the blade and are continuously pulverized before being blown back onto the lawn as nutrients. In a reel mower, pushing revolves an arrangement of cutting blades that slice the grass against a stationary blade. Most gasoline-driven mowers have a two- or four-stroke, one-cylinder engine with a connecting rod that converts the piston's to-and-fro motion into the crankshaft's rotary motion; a very basic carburetor is mounted outside. Handheld grass and hedge trimmers have blades that move side to side and back and forth respectively, their engines and cutting blades powered by household current or batteries. One popular type of weed trimmer cuts with a nylon line instead of a blade. At one end, a revolving cartridge rapidly spins short lengths of line that quickly slice through grass.

rotating blades

roller

Running on sun power, a robot mows a lawn (below). Mowers typically use gasoline engines or electric motors, both of which require careful operation to protect against injury, fire, or shock.

Heavy-duty equipment (opposite) makes large lawns seem smaller when blades of grass require cutting.

THE REEL THING: Its simple appearance notwithstanding, a push mower (left)—also known as a reel mower—can admirably tackle almost any lawn. This low-maintenance and nonpolluting machine also can work cardiovascular and other physiological wonders for home owners who use it regularly. Simply designed, the basic push mower holds a rotating cylinder of cutting blades, turned by a gear linked to the two wheels. As the blades turn, they cut the grass against a stationary blade between the wheels.

STRING TRIMMERS

Because many areas of a lawn may be beyond the reach of a conventional mower, a home owner may need other machines to help control grass growth or keep weeds in check. A string trimmer (left) uses a nylon cord instead of a metal blade to slash grass and weeds. At one end of the trimmer's long shaft is a gasoline engine or an electric motor; at the other is a spool that spins the cord, whipping its ends around more than 7,000 times a minute. Stretched tight by centrifugal force, the "strings" slice through light vegetation but remain flexible enough to bounce off tree trunks and fences.

LAWN SPRINKLERS

SEE ALSO

Cutting Edges
• 46

Smart Farming
• 120

IRRIGATION is the application of water by artificial means to crops, lawns, and gardens when weather and other environmental conditions make it essential. The practice dates back at least to the canals, locks, and reservoirs of ancient Egypt. On farms today, depending on the crop grown, water may be sent through fields in furrows or the fields may be flooded, as in rice field irrigation; it may come from sprinklers or through a variety of tubes on and under the ground. A home owner tending a lawn or a garden usually relies on an old hose with a hand-controlled nozzle, a network of plastic tubes that drip small quantities of water to the plants' roots, or an automatic sprinkler. Each option is based on the simple principle that as a fluid moves through a pipe, it exerts pressure on the walls that restrain it.

Lawn sprinklers have a broad range of designs. Some are merely doughnuts of compact metal or plastic tubing that have tiny holes punched into them; attached to a hose and moved manually from place to place, they simply spray out their fine arcs and curtains of water. Others spin or swing the water out of a punctured tube over a large area. They use a system of gears and a miniature turbine that essentially converts the kinetic energy from water coming through an attached hose to mechanical rotary energy, thus requiring no other source of power to move the spray tube back and forth.

Holes in a hose (left) offer one way to irrigate a lawn. Sprinkler irrigation saves water and lets the gardener select where the water should go.

Water sprite (opposite): A child delights in water that keeps a lawn green. Without regular watering by hoses and sprinklers, lawns would have to rely on uncertain rains.

AN OSCILLATING SPRINKLER (below) swings to and fro, sending up a curtain of water that drenches a wide swath of grass. It requires no power other than that supplied by the water flowing from a hose to a waterwheel. Gears turn a crank that shifts the rotary motion created by the passing water to oscillating motion, and the perforated sprinkler tube moves back and forth.

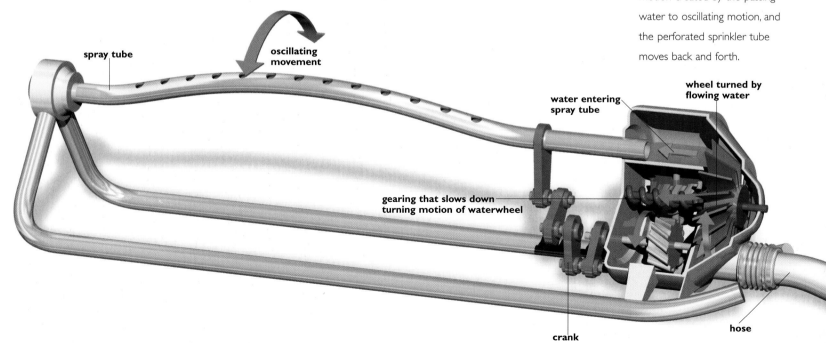

spray tube

oscillating movement

water entering spray tube

wheel turned by flowing water

gearing that slows down turning motion of waterwheel

crank

hose

POWER AND ENERGY

DEFINED as the capacity for doing work, energy takes many forms: It can be mechanical, electrical, physical, or thermal. When it is expended or dissipated by machines or human activity, we have power, which performs work. Sources of energy include water, wind, coal, oil, gas, wood, radioactive rocks, the moon's tidal pulls, and the sun's rays. To a significant extent, many people rely on biomass energy, employing waste-to-energy incinerators and other means to retrieve it. Coal and nuclear energy generate most of the electricity we use to light our buildings and power our machines. Waterpower supplies about a fifth of the world's electricity, while wind and geothermal energy are important sources of power for several countries. In the future, we may even look to exploding neutron stars for energy and, perhaps more immediately, to fusion—the same process that powers the sun.

Totems of power, high-tension wires transport the electrical power that runs nations.

SEE ALSO

King Coal • 184

Liquid Gold • 188

Wired • 30

IN 1886, William Stanley, a pioneer in the generation and transmission of electric power, fired up what was probably the earliest central power station. Erected in Great Barrington, Massachusetts, his station used a 25-horsepower boiler and engine to turn a 500-volt generator and supply electric lighting to 25 businesses. Stanley's creation and other early power stations relied on the same basic principles and components we use today, most notably fuel, boilers, turbines, and generators.

Coal is the fuel of choice at power stations. Before the 1973 oil embargo, the United States depended on oil to generate much of its electricity, but from 1973 to the 1990s, the use of oil dropped from 17 percent to 3 percent. Coal arrives at most stations ground into dust, which is then mixed with air to make a highly explosive fuel. Fired, it heats water in a boiler, thus producing high-pressure steam that pushes against the propeller-like blades of a giant turbine attached to the shaft of a huge generator. As the shaft spins, coils of wire interact with a circular array of magnets to create electricity.

Electricity then goes through a set of transformers, which are devices that alter voltage. First, they step up the voltage for main distribution along tall latticed towers. Later, at a substation, other transformers bring it back down to accommodate normal domestic voltage of 240 and 120 volts before it is distributed.

CURRENT: A simple AC generator (below left), which consists of magnets, a spinning coil with connecting slip rings, and carbon conductor brushes, uses electromagnetic induction to convert mechanical energy into electricity. As the coil spins and the induced current reverses in the magnetic field, the current passes from the rings to the external circuit by way of brushes that press against the rings.

In a giant turbine's maw (below), workers perform necessary maintenance. The blades of such machines must absorb the punishing effects of steam, hot gas, water, and air.

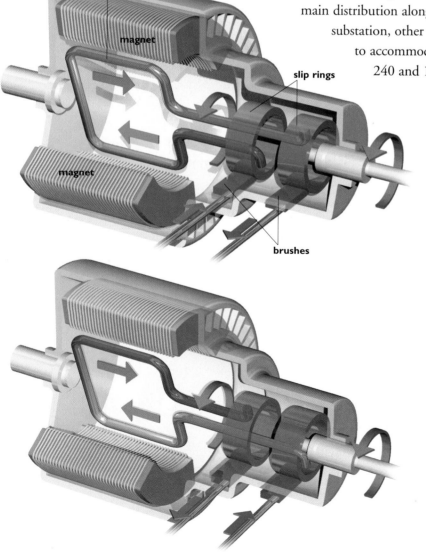

armature coil

magnet

slip rings

magnet

brushes

Power plants such as this one (left) in Tampa, Florida, help supply the world with electricity, its most versatile form of energy. While plants can use virtually any fuel to make electricity, about 55 percent of the electricity in the U.S. comes from coal-burning facilities. In the **control room** (below) of one plant, human and electronic eyes maintain a voltage vigilance. Because the distance that electricity can travel depends on the strength of the current, its flow must receive a boost to high voltages for main distribution. Substations use transformers to bring down the voltage to accommodate the normal domestic strengths of 240 and 120 volts.

WATERPOWER

SEE ALSO

Making Electricity • 54

Wired • 30

PLACE a waterwheel beneath a small waterfall. As it turns under the force of the water on its paddles, the energy can be shifted from its central, axle-like shaft to linked wheels, belts, and gears, and ultimately to useful tools. For centuries, people used waterpower to grind corn and wheat, to spin fiber, and even to fire up their charcoal-burning, metal-smelting furnaces. The Chinese, for example, are known to have employed a water-powered bellows in A.D. 31.

Waterpower is actually a misnomer because gravity is the heart of it, and water is only the medium that transmits gravity's action to a prime mover, such as a waterwheel. Waterpower—or hydroelectric power—is an efficient, economical way to harness the potential energy of falling or dammed water; indeed, hydroelectric power generates about a fifth of the world's electricity.

A hydroelectric power station requires a large head of water and a fall or gradient to take advantage of the water's gravity and momentum. For these reasons, such a power station is usually located in mountains, near a waterfall, or below a dam holding back the enormous force of a reservoir that results from blocking a river's flow. A hydroelectric power station is really a king-size version of the waterwheel, but its powerhouse contains generators and turbines whose blades spin under the carefully controlled force of a jet or flow of water.

Because of high costs and technical obstacles, one form of waterpower has potential but thus far little more than that: It would harness the enormous energy locked in the ebb and flow of the tides. Incoming tidewater would be trapped, run through turbines to generate electricity, and then released, letting it wash out to sea at low tide. Someone once calculated that the mean power dissipated by tides all around the world was about 1,100 million kilowatts; this figure has since been downsized considerably.

POWER STRUCTURE (below): A dam produces electricity, stores water for irrigation and public use, controls floodwater, and increases river depth for navigation. This massive barrier controls water by regulating its flow through gates and within long tunnels and pipelines called penstocks, or sluices. In a hydroelectric plant, water feeds through penstocks into the turbine. Bushings provide electrical insulation, and bus bars—lengths of conductors—collect the electric current and distribute it.

A bank of transformers (opposite) rises at McNary Dam on the Columbia River of the Pacific Northwest.

Outlet tubes in the Glen Canyon Dam (below) direct water from the Colorado River. Completed in 1966, the 710-foot-high dam supplies power and regulates the river's flow.

buttress

reservoir

gantry crane

bushing (insulator)

screen

generator

penstock

transformer

bus bar

afterbay

turbine

NUCLEAR POWER PLANTS

SEE ALSO

Fusion • 60

Making Electricity • 54

Tin Fish • 106

Wired • 30

A SINGLE ounce of the uranium-235 isotope can generate an immense amount of energy, a fact that makes it easy to understand why more than 400 nuclear plants in 30 countries—a quarter of them in the U.S.—rely on it to produce electricity. Nuclear energy now provides about 20 percent of U.S. electricity, and the distinctive domes and towers of nuclear plants have become sights as familiar as gas stations and shopping malls.

Each plant generates electricity with the same steam-turbine-generator arrangement used by a coal-fired plant. The fuel, however, consists of solid ceramic pellets containing isotopes of uranium atoms that are split apart in a process known as fission. The pellets are packed in long metal tubes, which are bundled together and installed in a heavily shielded and water-cooled reactor where fission occurs. When uncharged subatomic particles called neutrons are released, the particles collide with uranium atoms and split them. As the nuclei burst, the atoms release their own neutrons, which then strike other atoms to create a chain reaction; the immense heat generated by the process vaporizes water, creating steam that turns the turbine and generator. The power plant controls the reaction by inserting and removing rods made of neutron-absorbing material. Inserting these rods prevents neutrons from hitting atoms, and withdrawing them speeds up the reaction.

Letting off steam, the concrete cooling towers of a nuclear power plant (above) provide visible evidence of the enormous heat generated when atoms split. The actual waste from a reactor takes the form of a solid, not a gas, and although it will not explode, it still needs careful handling, usually by remote control and with adequate shielding.

A nuclear power plant (left) has a vaultlike interior designed to prevent the escape of radioactive material. Sensors adjust or shut down the reactor at the first sign of trouble.

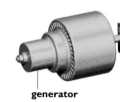

generator

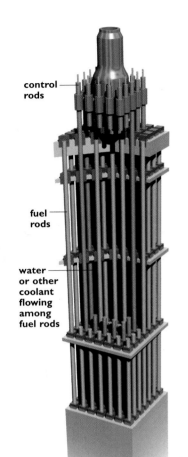

control rods

fuel rods

water or other coolant flowing among fuel rods

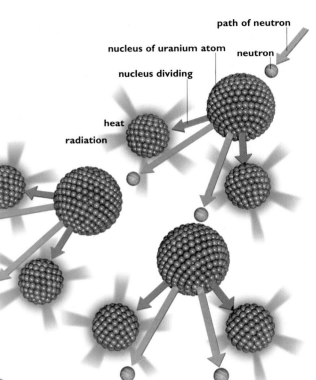

path of neutron

nucleus of uranium atom

neutron

nucleus dividing

heat

radiation

TWELVE-FOOT-LONG RODS (left) initiate and control a chain reaction. Shown in yellow, uranium fuel in ceramic pellets packs rods made of zirconium, a metal that resists heat, radiation, and corrosion. Control rods, shown in red, absorb neutrons, preventing them from hitting and splitting uranium atoms; inserting these rods stops the chain reaction, and withdrawing them speeds it up. Water flowing among the fuel rods carries off the heat.

A CHAIN REACTION (right) begins when neutrons smash into uranium atoms in the fuel pellets. The atoms fission, or split, releasing neutrons of their own. One fission spawns other fissions, which trigger more. As the atoms split, they liberate intense heat. In an atomic bomb, the reaction moves quickly, creating an explosion. In a reactor, though, the speed of the reaction is carefully controlled.

concrete

steel

steam generator

hot steam to turbines

concrete

turbines

reactor core

condensed water from turbines

coolant pump

NUCLEAR REACTOR: A fossil-fuel plant burns coal or oil to heat water in a boiler, turning it into steam that rotates a turbine. In the reactor process (left), fission heat turns water into steam that spins the turbine and the generator. But water also acts as a moderator. It removes heat from the chain reaction and slows down neutrons, increasing the probability of fission; it also helps to control the action, since loss of it slows or stops the reaction.

Unlimited Energy: Fusion

SEE ALSO

Alternatives
• 86

Fast Track
• 96

Nuclear Power
• 58

FUSION is the nuclear reaction that energizes not only the sun and the stars but also the hydrogen bomb, in an equally uncontrolled fashion. Unlike fission, which produces energy by splitting large, heavy atoms into smaller ones, fusion occurs when the nuclei of small, light atoms are compressed under intense heat to form larger and heavier nuclei, a reaction that also creates enormous bursts of energy. In theory, this energy could be captured and converted into electricity, a process that appears simple on paper. It is also appealing, because the fuel to start the process is plentiful: Deuterium and tritium, the heavy isotopes of hydrogen, can be extracted from ordinary seawater. Deuterium is a likely fuel candidate because only 1/250 of an ounce—the amount in one gallon of seawater—contains about the same energy as 300 gallons of gasoline.

The problem to be overcome is how to control such energy. Reactors would be required to withstand heat from plasma, the seething gas of charged particles resulting from a fusion reaction. Scientists have tested experimental containers that have no resemblance to a fission reactor's vault; they are instead made of coils of wire—"magnetic bottles" that create magnetic fields powerful enough to confine the manufactured "hell." The doughnut-shaped tokamak uses giant coils of superconducting materials in a riblike arrangement to generate a horizontal field that forces charged plasma particles to circle inside the container. The particles collide, fuse, and release energy without touching the chamber walls.

TOKAMAK (below): In 1951 the Soviet physicists Andrei Sakharov and Igor Tamm first proposed the tokamak, a device that contains hot plasma, fusion's charged gas. Inside the tokamak, which stands for "toroidal chamber with an axial magnetic field," superconducting materials create a magnetic field that causes plasma to flow within a doughnut-shaped chamber.

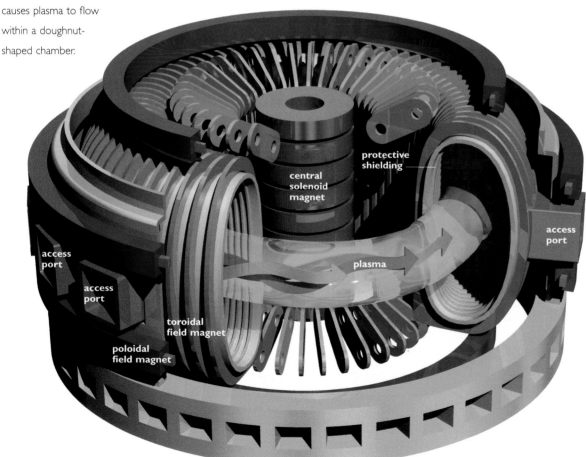

protective shielding

central solenoid magnet

access port

plasma

access port

toroidal field magnet

poloidal field magnet

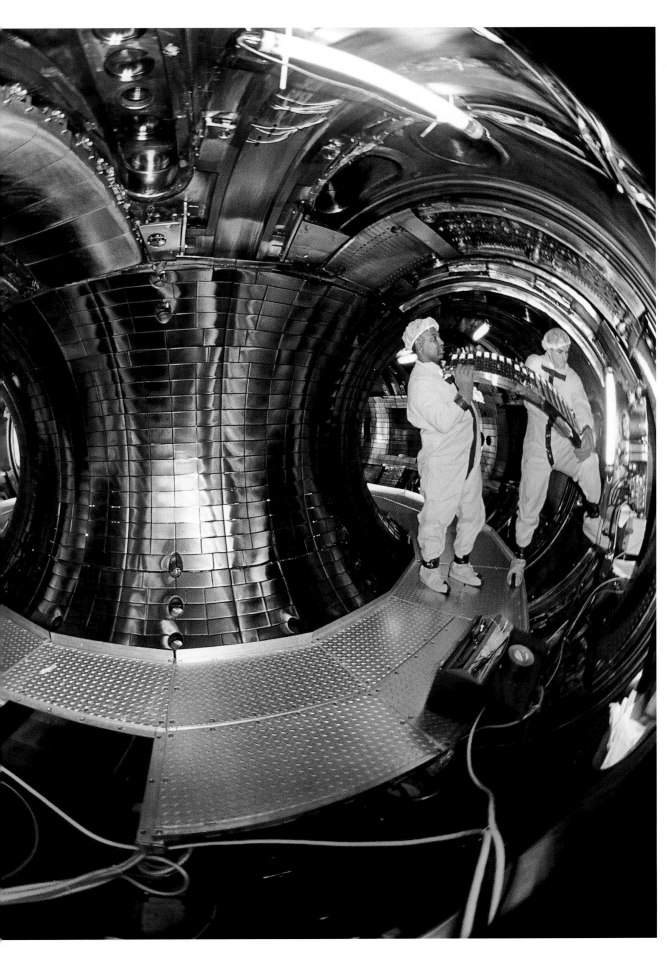

Belly of the beast:
Inside the tokamak, Princeton University researchers toil in an attempt to achieve the elusive break-even point at which energy released by fusion equals that required to produce it. Building plasma containers has proved a daunting task, given the furious heat they must hold. But with superconducting materials that contain heat more efficiently than copper, fusion power may live up to expectations as a source of safe and limitless energy.

WIND AND GEOTHERMAL POWER

SEE ALSO

Making Electricity
• 54

Solar Heating
• 26

ONE form of power captures the stream of air we know as the wind; the other taps naturally occurring steam and hot water below the surface of the ground. From time to time, both forms are painted as rather fanciful sources of energy, unrealistic answers to dwindling supplies of oil, coal, and gas. But where they are feasible, they work very well.

Like the waterwheels they resemble in principle, windmills have been used for centuries, their wind shafts connected to a series of gears and attachments that milled grain, irrigated fields, and pumped seawater from low-lying land. A wind turbine, the modern version of the windmill, produces electricity, driving a generator whose shaft-spin can be increased by gears; a computer may also control the rotor's movement and the position of the huge propeller-like rotor blades. The world's largest concentration of such turbines is in California.

Geothermal energy, in the form of hot water and steam, is the product of decaying radioactive elements in the ground. When tapped, it can drive a turbine to produce electricity, or it can be piped into buildings and used for space heating. In the U.S., geothermal power plants operate in California, Hawaii, and other states. Such plants are common in Iceland, where many homes are warmed by piped-in underground heat. Most underground heat is not exploitable, however, and it is concentrated in reservoirs found only in certain parts of the world. Often, the sources manifest themselves quite dramatically as geysers, steam vents, fumaroles, and the hot springs that have been used therapeutically for centuries. But surface displays may be far from a reservoir, and some sources of geothermal energy give no sign of their presence, requiring prospectors to use seismic and geologic probes that are complex and costly.

Hot water (far right): People relax in the bathlike waters near a geothermal power plant in Iceland, a country that, despite its name, has numerous volcanoes and hot springs. A tempting energy source, geothermal power cannot warm everyone: Only certain regions hold natural reservoirs of hot water, and even those sources run dry, may contain contaminants, and cost a great deal to develop.

Wind power (right): In Palm Springs, California, the wind helps generate electric power. Computers can adjust the propeller-like rotor blades on wind turbines to function in all wind conditions. To be effective, however, such windmills must stand where winds blow consistently.

TODAY'S construction projects often use wood, brick, stone, and concrete—the same materials ancient builders relied upon. To erect buildings of great size, complexity, and beauty, many projects also rely on iron, steel, aluminum, glass, and synthetic materials that were never dreamed of by early architects. Simplicity of design and function remain, but building today is an exact science, a major shift from the days when construction was a practical craft based often on experience and observation alone. Modern buildings, along with older ones that have been modified, are engineered to withstand winds, earthquakes, traffic, hurricanes, and fire, and most are probably far more comfortable, safe, and efficient than anything that was built in the past. The equipment used to erect these structures is also impressive. Occasionally, it is even more dazzling than the building it shapes.

Tubular tower of London: The steel frame of the Lloyd's building reaches skyward.

ELEMENTS OF CONSTRUCTION

SEE ALSO

High Steel • 70

Trade Tools • 68

Wood • 180

THE keystone is a simple, wedge-shaped block of stone on which much depends. Fitted last into the top of an arch, it locks the other pieces in place, providing support for the curved structure that, in turn, supports weight above it. Keystones and all of the other basic elements of construction work as a team to ensure stability and distribute weight. In constructing an arch, for example, wedge-shaped stones called voussoirs are placed on each side of the keystone. As gravity pulls the keystone downward, the thrust is carried on either side by the voussoirs immediately flanking it. The total thrust of the voussoirs is then distributed through the semicircle until it reaches the vertical part of the wall, which carries it directly to the structure's foundation.

Force applied to a building material will create internal stresses such as bending, compression, tension, and torsion (twisting), and to guard against

these stresses, structural forms abound, each with its own function. Natural base or prepared foundations support buildings ranging from sheds to skyscrapers. Columns support beams and lintels, horizontal spans that carry the load of a roof. Piers, arches, rib-vaulting, and buttresses, as well as flying buttresses and semidomes that absorb thrust, support huge domes.

Men at work (opposite): The geometrically precise framework of buildings under construction in California's Orange County has support, as well as design, as its goal. Horizontal sills of timber rest on the foundations; joists form horizontal frames for flooring; vertical studs provide support for walls; horizontal collar ties connect rafters on either side of roof ridgepoles.

Towering over Lisbon, Portugal, the Aguas Livres aqueduct (upper left) resembles structures built by the ancient Romans. Exemplars of form and function, their aqueducts arched across valleys and cities, carrying water from distant sources to baths and fountains.

BEARING THE WEIGHT (left): Columns, beams, an arch, and a foundation of steel-reinforced concrete provide interior and exterior support for a building. When carefully connected to one another, these important elements serve to distribute and support the great weight of a large structure.

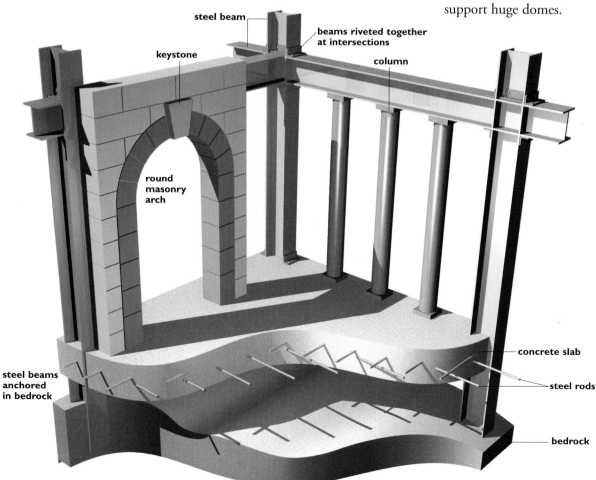

steel beam

keystone

beams riveted together at intersections

column

round masonry arch

steel beams anchored in bedrock

concrete slab

steel rods

bedrock

TOOLS OF THE TRADE

SEE ALSO

Construction Elements • 66

Cutting Edges • 46

High Steel • 70

ONCE, the main things that concerned a home builder were slope, drainage, and proximity to a water supply. Site preparation amounted to clearing away trees, underbrush, and rocks. Excavation, if any, was minimal, and a foundation was laid rather quickly with rough stones gathered from nearby fields. Using wood, bricks, mortar, and hand tools, the builder soon had the bearing walls up and the roof raised. Today's builders have other considerations: Faults can trigger earthquakes; soil can heave or sink structures when changes in weather or composition occur; high water tables can flood cellars; radon may be present in rocks. Tools are now nail-guns, power saws, routers, jackhammers, and electronic surveying devices. On a larger scale, they are dynamite, water pumps, and earth-moving bulldozers, graders, backhoes, and power shovels. A variety of cranes can work on the ground or atop skyscrapers.

Design, site preparation, and foundation laying must consider the strength of the supporting soil and rock. They must consider the "dead load" of the building—the total weight of the structure and all of its fixed equipment—and the changing "live load," which includes seismic forces, wind, and vibrations caused by equipment, people, and moving furniture. If subsurface conditions are adequate, foundation concrete is poured into an excavation. At other times, piles made of wood, concrete, or steel are driven down to firm soil or solid rock. Caissons, which are watertight cylinders filled with concrete to form columns, may also be sunk into weight-bearing soil that lies beneath fill or mud.

Jackhammer in hand, a worker (below) breaks up concrete, a strong building material made by combining sand, gravel, and rock with a binding cement of silica, alumina, and limestone.

Mixed with water, concrete's ingredients go from spreadable to moldable (above) and then harden into a rocklike material. Concrete withstands enormous stresses in highways, buildings, dams, and bridges when steel bars reinforce it.

Construction site (left): A work field for contractors and machinery, a site at first gives little hint of the splendid edifices that will rise above it.

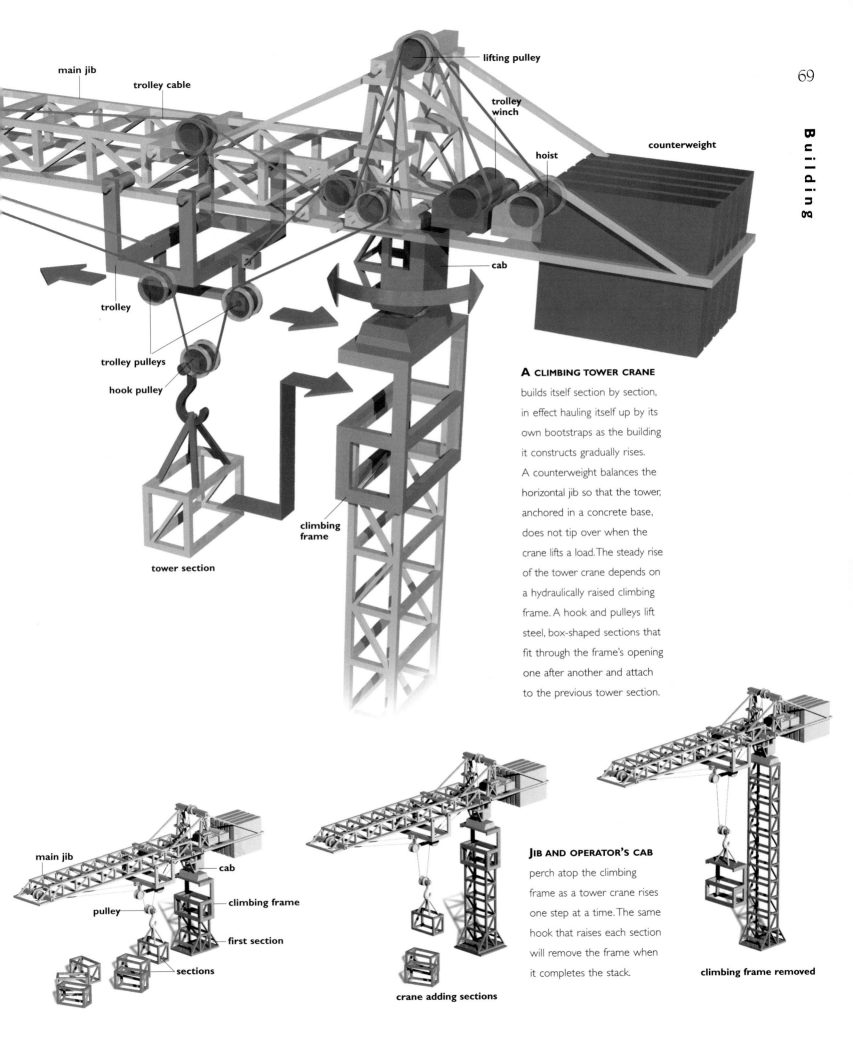

main jib

trolley cable

lifting pulley

trolley
winch

hoist

counterweight

cab

trolley

trolley pulleys

hook pulley

climbing
frame

tower section

A CLIMBING TOWER CRANE

builds itself section by section,
in effect hauling itself up by its
own bootstraps as the building
it constructs gradually rises.
A counterweight balances the
horizontal jib so that the tower,
anchored in a concrete base,
does not tip over when the
crane lifts a load. The steady rise
of the tower crane depends on
a hydraulically raised climbing
frame. A hook and pulleys lift
steel, box-shaped sections that
fit through the frame's opening
one after another and attach
to the previous tower section.

main jib

cab

pulley

climbing frame

first section

sections

JIB AND OPERATOR'S CAB

perch atop the climbing
frame as a tower crane rises
one step at a time. The same
hook that raises each section
will remove the frame when
it completes the stack.

climbing frame removed

crane adding sections

HIGH STEEL: THE SKYSCRAPER

SEE ALSO

Construction Elements • 66

Elevators • 74

Smart Buildings • 72

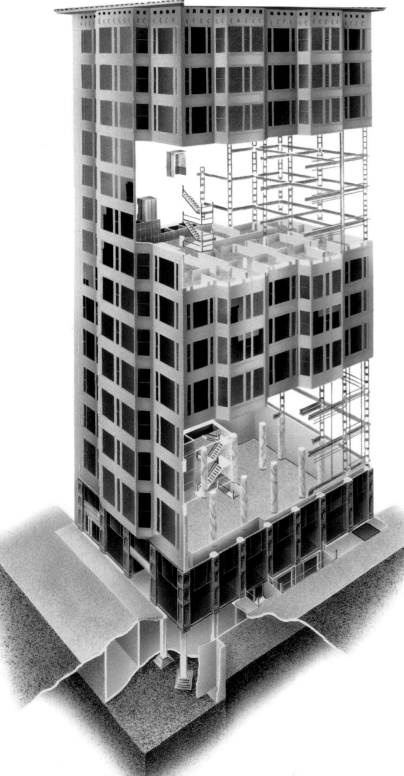

AN American original, the skyscraper began its gravity- and wind-defying rise in the late 19th century. It was a soaring testimonial to economic boom times, mass production, and the burgeoning technology of structural engineering. Developed first in Chicago, Carl Sandburg's "city of the big shoulders," the skyscraper relied on its internal steel skeleton of columns and beams, rather than conventional heavy masonry walls, to support most of its weight. A "curtain wall" made of nonbearing materials, such as glass and thin marble sheets, sheathed the framework, giving engineers and architects a relatively lightweight building with excellent tensile strength.

Innovations such as I-shaped steel beams, concrete reinforced with embedded steel, and tubular concrete designs allowing lighter and stronger walls to bear loads enabled skyscrapers to rise even higher. New designs provided more internal room by reducing the need for diagonal wind-bracing; prefabricated materials speeded construction; and improvements in concrete gave architects more freedom of expression.

Beyond aesthetics, efficiency, and the comfort of its occupants, a skyscraper is built to withstand high winds and other weather assaults, as well as the shaking of earthquakes. To protect against the lateral forces generated by earthquakes or winds, engineers today are constructing "nonrigid" buildings that, in effect, go with the flow or sway with the forces. One technique places layers of rubber and steel plates between a building's base and foundation. Another relies on sliding bearings under load-bearing columns, an arrangement that helps dissipate a quake's energy through friction.

CHICAGO'S 1895 RELIANCE BUILDING (left) reveals the network of horizontal and vertical beams supporting its 14 stories. Skyscrapers still rely on skeletons of steel to carry their enormous weight. But high-strength concrete now gives builders a strong option and more leeway when they design interiors and exteriors. Used in a popular design that transforms a building's exterior walls into a rigid tube, concrete carries gravity and wind loads. The interior of such a building allows more innovative and practical design because the structure itself needs fewer internal beams, girders, and columns to bear its weight.

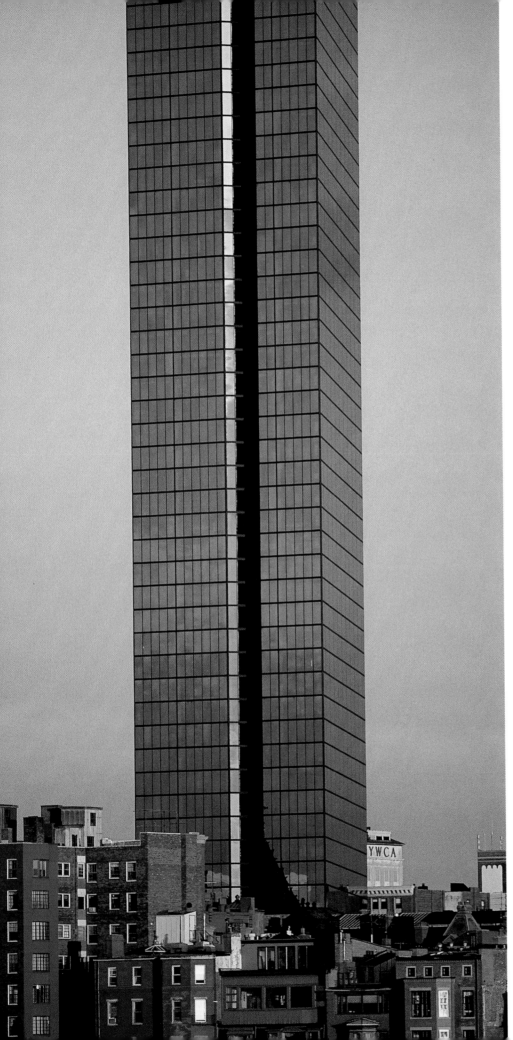

Raised and praised to the skies, a glass-sheathed, sun-splashed high-rise (left) humbles the structures that cluster at its feet. Such towers symbolize the engineering prowess, imagination, and audacity of humankind.

Building a high-rise (above) requires the vision of architects, acres of urban land, and a king's ransom. It also needs the skill of steelworkers, stonemasons, glaziers, welders, and countless other workers seemingly oblivious to great heights.

SMART BUILDINGS

SEE ALSO

E-Mail • 244

High Steel • 70

On Call • 218

Sensors • 42

SKYSCRAPERS and other large buildings are more than walls, partitions, and foundations. From air-conditioning and heating ducts to mail chutes that drop letters to a central pickup point, buildings are laced with familiar systems for environmental control, transportation, communication, power, water, and waste disposal.

With new, computerized technology, "intelligent" buildings can respond quickly to the needs of occupants. Electronically enhanced and regulated by central computers, buildings are wired to respond to many contingencies. Fire alarm and security systems can be linked so that doors can unlock automatically to provide the quickest evacuation route, while dampers regulate airflow to inhibit the spread of flames. Sensors in sheets of "smart" composite materials detect leaking gas and alert engineers when repair is needed. Whole walls studded with actuators are transformed into speakers, and some systems monitor and correct other systems. From a central control station—not necessarily in the same building but perhaps in another structure miles away—a building manager can turn out lights, limit power consumption, run boilers more efficiently, and stop elevators in an emergency.

Workplace automation that relies on shared equipment and systems to integrate the data stream has contributed greatly to the efficiency of companies and offices. Among a smart building's many tools are centralized data and word processing, teleconferencing and electronic mail systems, and message centers that answer the telephones automatically.

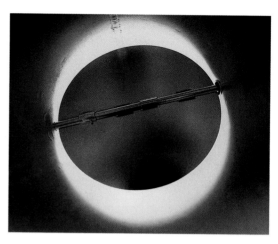

Basic in design but with a large job to perform, the volume regulator (above) of a climate control system dictates whether a building warms or cools. To activate it, a person only needs to twist a dial on a thermostat.

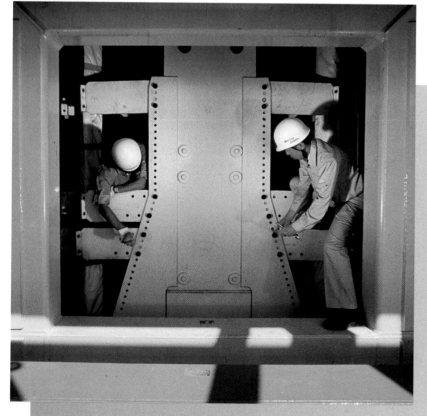

EARTHQUAKES

Earthquakes can devastate rigidly constructed buildings, but damage-control technology largely attributed to Japanese and U.S. engineers has made structures less vulnerable to nature's assaults. Using machines that simulate earthquakes, scientists study the effects of quakes on tall buildings, bridges, nuclear power plants, gas tanks, oil pipelines, and even household equipment such as refrigerators and ranges. What they learn helps engineers build structures that absorb and dissipate a quake's destructive energy. The Osaka World Trade Center Building in Tokyo (left), for example, uses a computer-directed sliding weight to shift the structure's center of gravity when the Earth trembles or the wind rises. To dampen shock, another method uses alternating layers of steel and rubber between a building's base and foundation.

Changing conditions:
Environmental control in a
mixing box, the Mischbox
in Germany's Aachen Clinic
exemplifies the intricate
technology behind heating,
cooling, ventilation, and
humidity control. Increasingly
automated, climate control
systems rely on computer
chips and localized sensors
that respond quickly to shifting
conditions. Such systems
have many desirable aspects.
Without humidity control, for
example, human skin dries,
sinuses clog, furniture shrinks,
hardwood floors split at the
seams, pianos go out of tune,
and wallpaper peels.

ELEVATORS AND ESCALATORS

SEE ALSO

High Steel • 70

**Smart Buildings
• 72**

PERHAPS the most essential piece of technology in a high-rise building is the people mover. Indeed, elevators and escalators made skyscrapers possible. They have had extraordinary economic impact on businesses and public facilities, and they are factors to be reckoned with when architects design new buildings or transportation centers. Both are "lifts" that require motors to haul them up and down, but the similarity ends there.

The first elevator was developed in the middle of the 19th century by Elisha Otis, an American inventor. Powered by steam, his device ran up and down between guide rails and had an automatic safety device that clamped on the rails to prevent the car from falling if the hoisting rope broke. Hydraulic systems that were run by water pressure powered later elevators. Some small buildings still have them, but most of today's models operate with steel hoist cables drawn up by motor over a grooved pulley wheel, called a sheave. The cables are attached to a heavy counterweight that goes up when the car descends, thus balancing the weight of the car.

Escalators are moving stairs mounted on a continuous chain that is drawn up over a drive wheel run by an electric motor. The stairs form a sort of level treadmill at the beginning and end of the ascent, becoming an arrangement of treads and risers during the incline. While elevators are swifter and can fit into small wells inside a building or run up its sides, they carry a limited number of passengers; escalators do not stop, and they carry more people relatively quickly.

Modern design (below): A tableau in the Market Tower Building of Frankfurt, Germany, suggests that the elevators of today—glitzy descendants of crude lifting devices once used in mines and warehouses—not only provide vertical transport but also serve as important elements in a building's design. Hoisted by clanking, steam-run machinery several decades ago, elevators now glide swiftly and quietly within high-rises or even along building exteriors.

THE LIFT (far right): What goes up goes down, and vice versa. The elevator, or "lift," follows this simple principle. Without elevators, people would have long walks, and skyscrapers would be impractical.

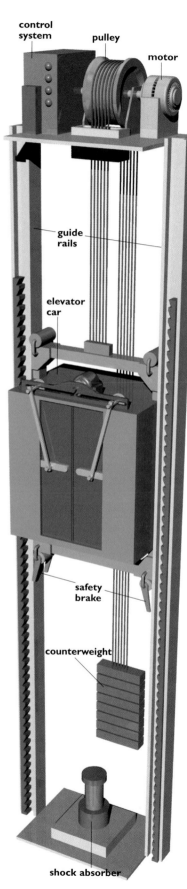

control system

pulley

motor

guide rails

elevator car

safety brake

counterweight

shock absorber

Paths cross (left) as shoppers enjoy the ease of an escalator network in a Marshall Field's department store.

INCLINED ROUTE (below): Up stairs become down stairs, and handrails appear and disappear as an escalator moves effortlessly over its inclined path. Built on the endless chain principle, an escalator relies on an electric motor at the top. A chain— drawn around a drive wheel at the top of the landing and a return wheel at the bottom— moves the treadmill-like stairs attached to it. As the stairs move upward, they form treads and risers that flatten at the end of the ascent and again at the beginning.

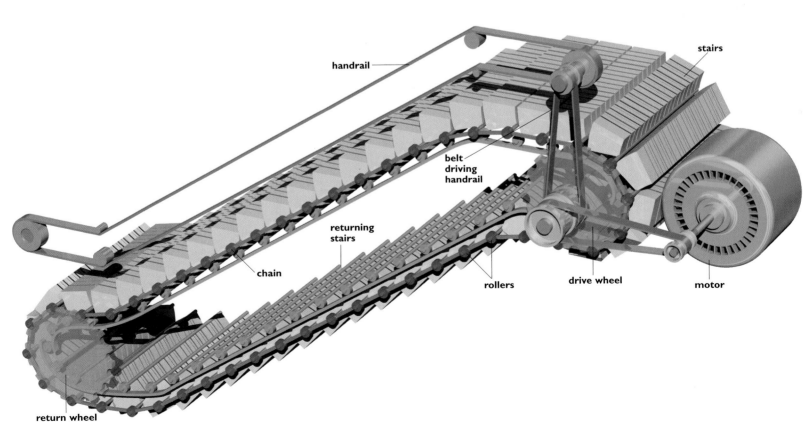

handrail

stairs

belt driving handrail

returning stairs

chain

rollers

drive wheel

motor

return wheel

SEE ALSO

On, Over, Under Streets
• 94

Trade Tools
• 68

Tunnel-boring machine
(opposite): One of eleven mammoth boring machines dwarfs workers on the Channel Tunnel project. The machines' rotating drill heads cut through rock 200 feet below the floor of the English Channel.

DIGGING a tunnel involves a slow, seemingly blind, passage of equipment and workers underground—through a mountain, under streets, or beneath a riverbed—with the crews and machines excavating, blasting, and reinforcing their way from the two opposite ends until they meet.

Tunneling requires heavy cutting equipment, such as huge tunnel-boring machines (TBMs) that have rotating drill heads. Moved slowly along by gigantic hydraulic rams and guided by computer-linked lasers and other electronic distance-measuring devices, these engineering marvels simultaneously cut through and eject tons of soil and rock. They also line the tunnel with prefabricated concrete or cast-iron segments.

In the most expensive privately financed engineering project in history, TBMs completed the 31-mile-long, three-tunnel "Chunnel" linking England to France. The tunnel's 24-mile underwater section, burrowed into a chalk layer well beneath the English Channel, is the longest in the world. More than just a tunnel, it has two one-way rail tubes for shuttles, freight trains, and high-speed passenger trains. The Chunnel's central service tunnel provides fresh air and allows for emergency evacuation and maintenance.

Designed to bore smooth walls to the exact size desired, TBMs eliminate blasting accidents and deafening noise. They are very expensive, though, and an alternative is to lay precast tunnel sections into trenches where they are joined and backfilled. Despite the high-tech safety precautions of either method, tunneling remains one of the most dangerous engineering jobs.

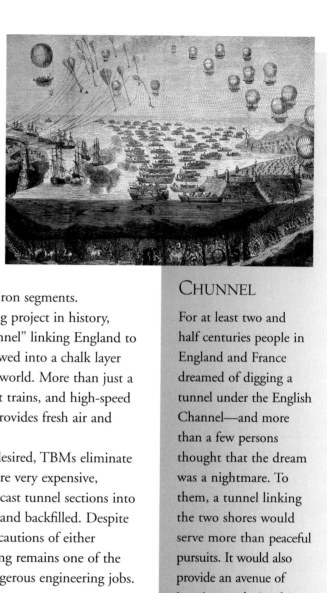

CHUNNEL

For at least two and half centuries people in England and France dreamed of digging a tunnel under the English Channel—and more than a few persons thought that the dream was a nightmare. To them, a tunnel linking the two shores would serve more than peaceful pursuits. It would also provide an avenue of invasion, as depicted in a fanciful 1803 engraving (above) that showed the French Army taking the low road into England.

ventilation

cross passage

piston relief ducts
(positioned every 820 feet)

train

service car

crossover
cavern

service tunnel

main tunnel

service car

main tunnel

lining segments

CHANNEL TUNNEL:
Embedded in a chalk layer, the "Chunnel" contains two one-way rail tunnels for trains and shuttles and a central service tunnel that provides ventilation, room for maintenance, and an evacuation route. Crossover caverns allow trains to change tracks, while piston relief ducts balance pressure waves caused by fast-moving traffic. Cross passages carry air from service tunnel to main tunnels.

A MUDDY BUSINESS: DREDGING

SEE ALSO

Eyes and Ears
• 220

Trade Tools
• 68

Waterpower
• 56

SCOOPING or sucking muck and mud from the bottom of harbors, rivers, or lakes may not sound like activities in which technology would dirty its hands. The fact is, however, that dredging relies heavily on sophisticated hydrographic equipment, on satellite information to accurately position the dredges, and on hydraulic systems. It also uses giant trailing-hopper dredges that hold enough soil to construct an entire island.

The dredging business is inextricably linked to the development and fate of many regions of the world. Dredges maintain and build dikes, dams, and waterways. They help construct canals, harbors, water-supply systems, and bridges. They also reclaim submerged land, mine for minerals and precious metals, and scour the ocean bottom for specimens of marine life.

Dredging is either mechanical or hydraulic, and the composition of the soil beneath the water dictates which method is used. Human bucket brigades and horse-driven bucket dredges were the mechanical means of excavation used as far back as 1600 in Amsterdam. Today's mechanical dredges use movable arms to scoop up mud in buckets and deposit it on barges or into hoppers aboard the dredging vessels. Grab-bucket dredges lowered from booms may be used for deepwater dredging; ladder dredges move buckets over an endless chain.

In hydraulic dredging, which was developed in the late 19th century, soft underwater material called spoil is loosened by revolving cutters or water jets, suctioned into a pipe, and discharged on shore through a floating pipeline.

Big dig (opposite): Deep and dirty work, dredging a channel requires huge vessels and barges and, on occasion, some of the more familiar tools of the construction trade.

Into the hopper: Dredges often float on watertight hulls of various design, but the job of each one involves excavation of underwater soil. The split-hull hopper dredge (left) dispenses with the need for barges to cart away the sludge, placing it instead in hoppers, or holds. When the hoppers reach capacity, the vessel leaves the site, opens the hopper doors, and disgorges the material.

EAGLE 1

NEW ORLEANS LA.

THE MOVE

WITHOUT fuel and engines, wind and sails, or arms and oars, all our vehicles would remain as motionless as bowling balls waiting to be bowled. We know this is true, but most of us cannot explain why. It took Sir Isaac Newton (1642-1727) to shed light on it all: "Every body perseveres in its state of rest or of uniform motion in a straight line unless it is compelled to change that state by forces impressed thereon." His other laws of motion tell us that acceleration depends on the amount of force exerted and that every action has an equal and opposite reaction. The forces instigating change may be physical, mechanical, electrical, magnetic, or gravitational, but no matter which furnishes the drive, the goal is the same: a change of position. While the concept is simple, it and all of its ramifications nonetheless laid the foundation for modern science and for all of our means of transportation.

Rail power: A fast-moving train streaks through the Mojave Desert in southern California.

SEE ALSO

Alternatives
● 86

Assembly Line
● 84

Sky Telephone
● 216

OTHER than a home, an automobile is generally the largest single investment that the average consumer makes. A hybrid from the Greek *autos*—self—and the Latin *mobilis*—movable—the word "automobile" was once simply an adjective that meant "self-moving." Today, the word is also a noun that indicates a powerful, self-propelled, passenger-carrying vehicle whose omnipresence and relative ease of operation often hide the fact that it is an awesome example of advanced technology. The modern automobile is a far cry from Leonardo da Vinci's 15th-century concept of a steam-propelled vehicle, and it's a long way from the three-wheeled, three-mile-an-hour steam carriage that was first built in 1769 by Nicolas-Joseph Cugnot, a French army officer. It is both a designer's

dream of beauty and a model of efficiency for engineers and theoreticians alike.

While the standard car's basic operation has changed little over the years—the gasoline-fueled internal combustion engine is still its source of power—the modern automobile is a paradigm of technological progress. Electronic ignition greatly extends the life of spark plugs; front-wheel and four-wheel drives help provide more traction; and a fuel-injection system helps control the air-fuel mixture to the engine. It also has superchargers for more power, antilock disc brakes that self-adjust for skids, emission controls that are designed to reduce the amount of pollutants released into the environment, and enough advanced, computer-controlled systems for convenience and performance to thrill any technophile.

The flimsy gasoline carriages invented in the late 19th century, including Henry Ford's autocycle and Charles Duryea's buggyaut, seemed to consist only of a seat, a steering tiller, and an engine bolted between bicycle or wooden wheels with iron fittings. A cutaway illustration of one of today's automobiles, revealing components that were not even dreamed of in the early days of the horseless carriage, would probably baffle the early inventors far more than it would modern drivers. This generation is rapidly becoming accustomed to such innovations as navigation systems that provide automatic, voice-guided route selection.

An air bag (below) inflates with explosive force against a test dummy when a strong impact ignites a detonator cap. Air bags save lives, but they also can cause serious injury, even death, in rare instances. New models that inflate with less force reduce the risk.

Computerized navigation (above) comes to the aid of a baffled traveler on the road. Installed in a rental vehicle, an electronic navigation system maps a car's precise position on a highway, identifies routes, and selects the best way to the driver's destination.

Reinvented wheel (above): Running on electric power, the 1932 Dynasphere revolves around a stationary driver.

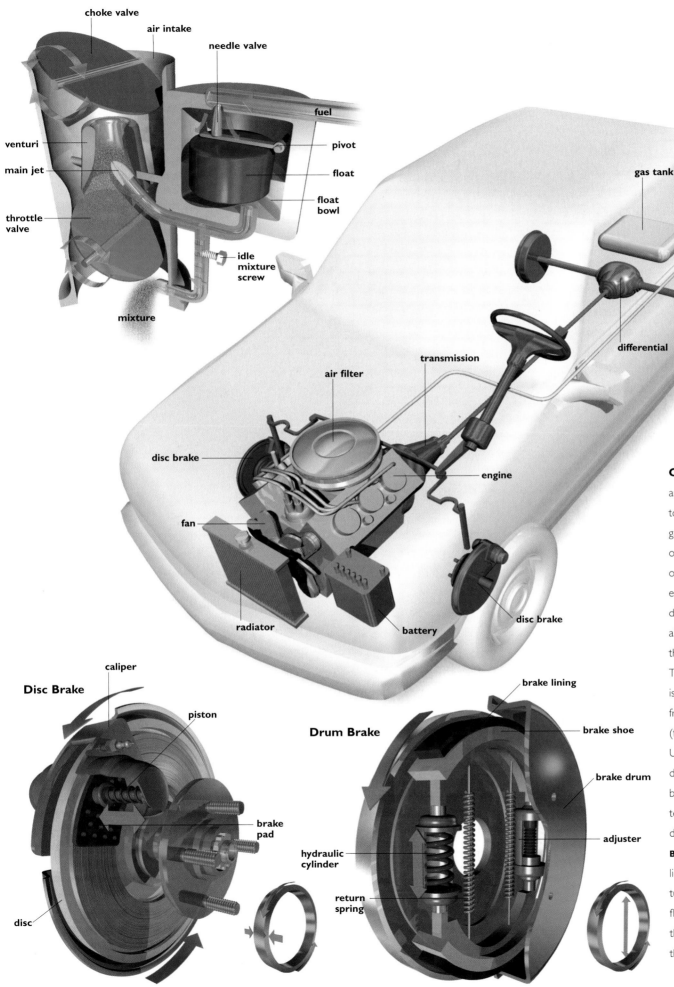

choke valve

air intake

needle valve

fuel

venturi

pivot

main jet

float

throttle valve

float bowl

idle mixture screw

mixture

gas tank

differential

drum brake

transmission

air filter

disc brake

engine

fan

radiator

battery

disc brake

Disc Brake

caliper

piston

brake pad

disc

Drum Brake

brake lining

brake shoe

brake drum

adjuster

hydraulic cylinder

return spring

CARBURETOR (top left) and **ENGINE** (left) team up to ignite and burn vaporized gasoline inside the cylinders of an automobile—the basis of the internal combustion engine's power. The carburetor delivers the correct fuel and air mixture to the cylinders through an intake manifold. The hydraulic braking system is equally sophisticated. In front-wheel **DISC BRAKES** (far left), friction pads inside U-shaped calipers straddle discs that turn with the wheels; brake fluid forces the calipers to press the pads against the discs. In rear-wheel **DRUM BRAKES** (left), stationary shoes lie inside cupped drums that turn with the wheels; brake fluid forces the shoes against the drum surfaces, slowing the automobile.

SEE ALSO

Alternatives
• 86

Automobile
• 82

Automobile bodies

(below) parade past automated machinery that assembles and fits them with parts from numerous warehouses and factories.

"THE motor car is not the product of a single inventor, nor even of men within a single century," wrote Charles F. Kettering, inventor and legendary head of research for General Motors Corporation. Written in 1929, his words still hold true. Neither the design nor the manufacture of a new car is the work of a handful of people, let alone one engineer.

Early automobiles, some electrically driven, were produced in small numbers just before the start of the 20th century. Many of the models were made in Europe, and only a wealthy few owned them. The Duryea Motor Wagon Company, founded in 1895 in Springfield, Massachusetts, built 13 cars that year. Three years later, the entire United States held fewer than a hundred gasoline-operated cars. Understandably, consumer demand was not high. Manufacturing facilities were limited, roads were poor, gas supplies were low, and cars were unreliable and costly. Today, 190 million passenger cars are in operation in the U.S., a testimonial not only to the overwhelming demand for comfortable,

Autoworkers (above), unhindered by doors and windows that special machines will add later, install essential parts and systems in the suspended space frame of a nearly completed car.

Robotic arms (below) precision weld thousands of spots on auto bodies, causing sparks to fly over a Honda assembly line in Japan.

reliable, and economical transportation, but also to mass production and a revolutionary manufacturing process: the assembly line.

Two Americans share credit for supplying the demand. Beginning in 1901, Ransom Eli Olds, the father of automotive mass production and Olds Motor Works, put together cars with engines and transmissions made in other shops. His Oldsmobile was the first commercially successful U.S. car, but the car that captured the popular imagination was Henry Ford's lightweight yet strongly built Model T, first manufactured in 1908. The affordable, easy-to-maintain Tin Lizzie was the first automobile built by modern mass-production methods. Engine and chassis assemblies were tested on one level and driven to the bottom of a chute; there, bodies were slid down and bolted on, and the cars were driven off.

Today, an assembly line is a round-the-clock operation that may reach ten miles in length. Along the way, giant presses shape and stamp out parts, and robots weld the automobile frame. The body is dipped into a rust-proofing bath, and coats of paint are baked on. Some machines fit doors and windows; others lower engine and transmission into the frame. From start to finish, the process takes 22 hours.

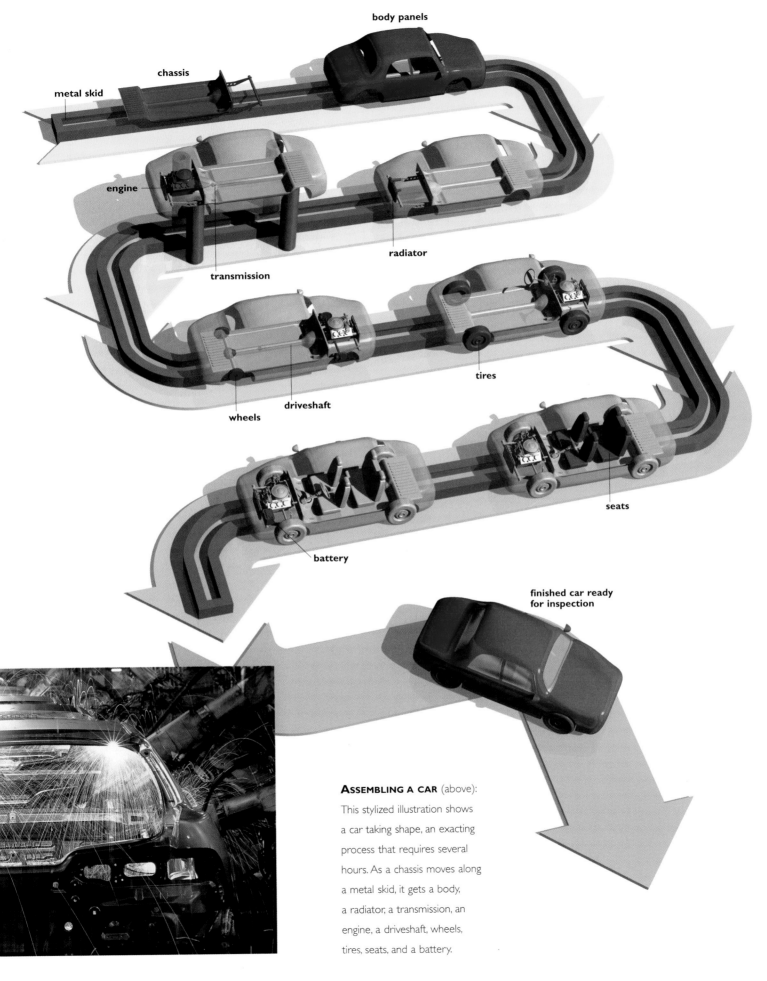

body panels

chassis

metal skid

engine

transmission

radiator

wheels

driveshaft

tires

battery

seats

finished car ready
for inspection

ASSEMBLING A CAR (above):
This stylized illustration shows
a car taking shape, an exacting
process that requires several
hours. As a chassis moves along
a metal skid, it gets a body,
a radiator, a transmission, an
engine, a driveshaft, wheels,
tires, seats, and a battery.

ALTERNATIVES TO THE PUMP

SEE ALSO

Automobile
• 82

Liquid Gold
• 188

**Making
Electricity**
• 54

SOME immediate and not-so-immediate concerns are causing automotive engineers to focus more and more on alternatives to gasoline. One of today's major concerns is air pollution caused by gasoline-powered automobiles. Another worry relates to the future: Our limited fossil fuel supply will not last forever. To meet our needs, now and tomorrow, engineers are again looking to electricity.

The ordinary, gasoline-powered car already relies on electricity—up to a point. Its electrical system furnishes electricity to operate the starter, ignition system, and accessories, and it recharges the storage batteries. But this kind of automobile also needs pistons and a carburetor, water pump, and muffler, none of which are found in an electric car. A simpler machine, an electric car sends electrical energy stored in the battery directly to the motor, where it is converted into mechanical energy. An onboard recharger plugs into an outside electrical outlet for "refueling."

While owners of electric cars don't have to deal with oil changes, they do have to put up with some disadvantages: Their vehicles are relatively slow and use heavy lead-acid batteries that take two or three hours to recharge after only about a hundred miles of road time. The answer may be new batteries of nickel-metal hydride, sodium-sulfur, or lithium, or possibly batteries that store solar energy.

Engineers are also testing hybrid automobiles that use fuel cells or electric motors in combination with gasoline engines. A fuel cell works like a battery except that it does not need recharging. It is basically a pair of electrodes wrapped around an electrolyte, an arrangement that generates heat and electricity by combining oxygen and hydrogen without combustion. The hydrogen fuel powering the cell is actually separated from the car's gasoline, but the fuel cell utilizes so much more of the gasoline's potential energy than a piston engine does that a car's range may be doubled on the same amount of gas, with less pollution. Even more environment-friendly is hydrogen that does not have to be extracted from fossil fuel; it can be removed from water by sunlight, a feat already accomplished by scientists.

Hybrid cars work in various ways. Some use the gasoline engine to charge batteries running the electric motor. Others switch between engine and motor according to driving conditions. A car developed in Japan starts on battery power and runs on a combination of gasoline and electric power when it speeds up.

Ecomobile (above): Part motorcycle and part car, this vehicle offers a quiet, reliable means of transportation that produces little pollution and goes without maintenance for long periods of time.

An electric motor (above) needs fewer moving parts than a conventional car engine. This model of simplicity has electrical or solid-state components but no complex emission-control system.

Fueling an electric car (below) requires only connecting the onboard battery to an outlet. On the downside, the heavy batteries take hours to recharge and run out after about a hundred miles.

In a sleek electric car (left), a technician testing the horsepower under the hood acts more like an electrician running a circuit check than an auto mechanic.

HYBRID POWER (below, from left to right): A gasoline engine splits power between wheels and generator, running the motor and recharging the battery. As the engine shuts off, energy flows from the wheels through the motor to be stored in the battery. When the car accelerates, the battery's energy assists the drive power.

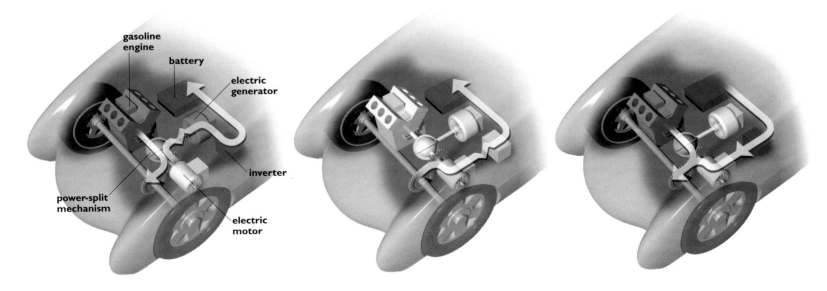

gasoline engine

battery

electric generator

power-split mechanism

inverter

electric motor

WHEEL POWER

SEE ALSO

Automobile
• 82

Small Appliances
• 20

RIDING a bicycle is one of the simplest ways to travel: It generally requires a minimum of expertise beyond knowing how to shift gears and keep one's balance. But the bicycle, which dates back more than 200 years, is an invention whose relatively humble appearance belies its complexity.

The early bike had iron tires that were first propelled by the rider's feet, then by ropes wound around an axle and hitched to a lever for driving power. The seat was uncomfortable, to say the least, because it was usually no more than a wooden beam. Even with refinements, the first bikes were frequently ridiculed: One 18th-century model was dubbed "dandy horse" by Englishmen who thought the frame resembled a pack animal. A modern bicycle, however, can be a high-tech, designer dream-machine equipped with a score of gears, an aerodynamic configuration, airfoil tubing, and a titanium cantilever frame.

In a sense, the rider is a bicycle's engine, supplying the physical energy to be converted to mechanical. The bike's chain serves to transmit force from one place to another, in this case from the pedal sprocket to the rear driving wheel. Gears on the front sprocket and the rear wheel serve the same purpose as other gears: They change one rate of rotation to another. This makes the bicycle very efficient, because the effort applied to the pedals can be geared up for high speed or geared down for hill-climbing power. A pair of devices called derailleurs transfers the chain from one sprocket to another. In highest gear, the rear wheel turns many times for each turn of the pedals, and the bike moves along swiftly; but in low gear, when a stronger forward shove is required to go, say, up a hill, the wheel turns fewer times in relation to the pedal, thereby trading speed for ease of pedaling.

Fill 'er up (above): A 1960s rocket roller, wearing a one-horsepower engine on his back to power his skates, refuels at a service station.

On the ice (left), a bike racer garbed like a space traveler leans on modern bicycle technology to break a world speed record.

MOTORCYCLES

An American legend in daring purple, a Harley-Davidson motorcycle exudes bravado even when parked. Able to accelerate quickly, a motorcycle has a two- or four-stroke gasoline engine with up to six cylinders mounted between the wheels. A foot-operated crank or an electric starter gets the machine up and running, while disc brakes— activated by a foot pedal for the rear wheel and by a hand lever for the front—slow and stop a finely engineered bike. Used by the police, by the military, and for recreation and travel, motorcycles were hailed as an art form in a 1998 exhibit at New York's Guggenheim Museum.

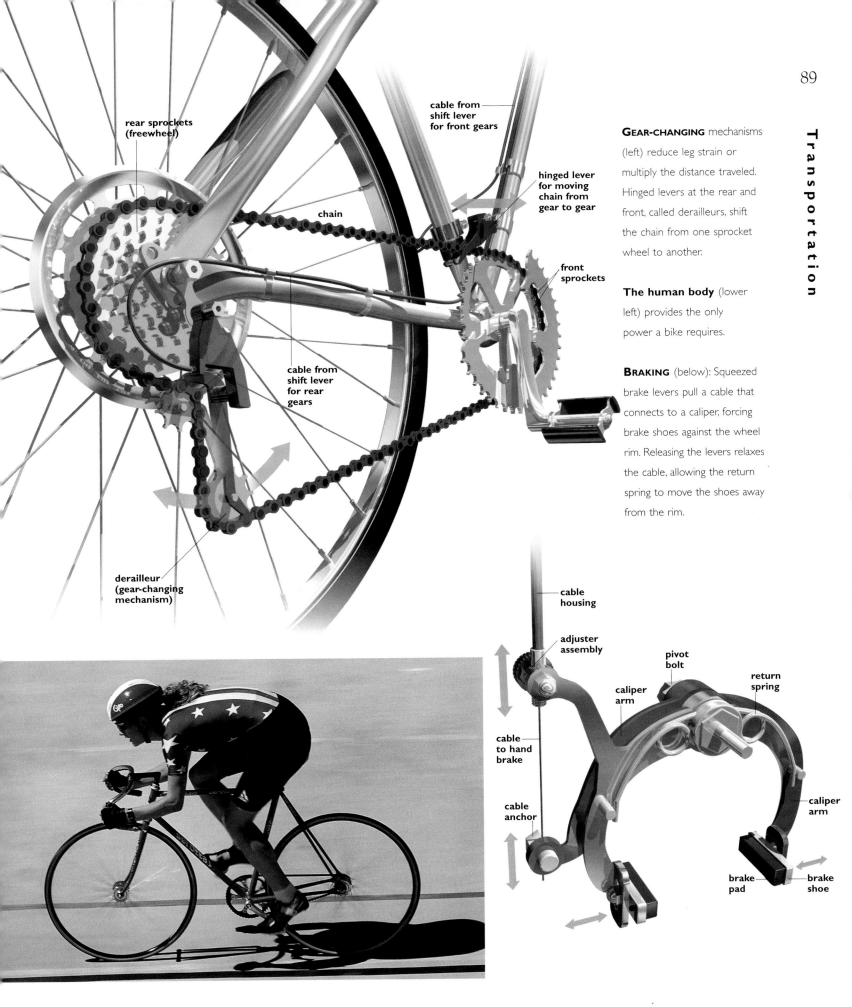

rear sprockets
(freewheel)

cable from
shift lever
for front gears

chain

hinged lever
for moving
chain from
gear to gear

front
sprockets

cable from
shift lever
for rear
gears

derailleur
(gear-changing
mechanism)

GEAR-CHANGING mechanisms (left) reduce leg strain or multiply the distance traveled. Hinged levers at the rear and front, called derailleurs, shift the chain from one sprocket wheel to another.

The human body (lower left) provides the only power a bike requires.

BRAKING (below): Squeezed brake levers pull a cable that connects to a caliper, forcing brake shoes against the wheel rim. Releasing the levers relaxes the cable, allowing the return spring to move the shoes away from the rim.

cable
housing

adjuster
assembly

pivot
bolt

return
spring

caliper
arm

cable
to hand
brake

cable
anchor

caliper
arm

brake
pad

brake
shoe

SEE ALSO

Diesel • 92

Fast Track • 96

Smart Trains • 98

"TRAINS are wonderful," the mystery writer Agatha Christie once observed. "To travel by train is to see nature and human beings, towns and churches and rivers, in fact, to see life." As countless other train-lovers know, moving people over the ground does not always necessitate rubber tires hitting the road. Steel wheels on steel tracks do the job admirably, and the development of the steam locomotive in the early 19th century proved that even an inefficient piece of equipment can suffice.

Its iron horse image and "King of the Railways" title notwithstanding, a steam locomotive was a coal stoker's backbreaking nightmare, a noisy machine that wasted energy, dirtied the air with soot, and cost a great deal of money to operate. One such juggernaut, which ran at a mile a minute in Rhode Island and Connecticut in the early 1900s, burned 3 tons of coal before pulling out of the roundhouse, carried 4 tons on a tender, and required 4,000 gallons of water in its tank. Revered in ballad and folklore, and romanticized by model railroaders and preservation societies, the world's remaining steam locomotives puff along these days in only a few countries. They also are found on old railroad lines that have been turned into museums, where even the T-rails and the wooden ties over which the antique engines still travel can capture our attention.

Admittedly, there is something about a chugging, huffing steam locomotive hauling a line of clickety-clacking coaches and boxcars, red caboose trailing in back, that stirs nostalgic souls. But something more stirring may be found in the technology behind the thing: the enormous piston-pushing power of steam produced at high pressure. With its piston rods connected to driving wheels as much as 85 inches in diameter—an arrangement akin to a tricycle's pedal wheel—the steam locomotive is the perfect visible example of how a machine can convert one form of movement into another. As the driving rods move back and forth under the force of steam, their linear movement is shifted into rotary mode, turning the wheels. The legendary John Henry may have wanted to die with a hammer in his hand rather than let a steam drill beat him down, but it wasn't worth the effort, technologically speaking.

Belching smoke, a steam locomotive (left) arrives at Portola, California. Once rulers of America's rail lines, steam engines and their intimidating bulk gave way to more refined, and more efficient, diesels and electrics.

Control elements (opposite, bottom left) The collection of safety valves, pumps, gauges, and hot tubes in Argentina's Old Patagonia Express presents a challenge even for a steamfitter.

Glowing steam gauges (opposite, bottom right) flank an engine driver who works on a Munich rail line. Perhaps he ponders the romantic past and the not-so-promising future of his iron horse.

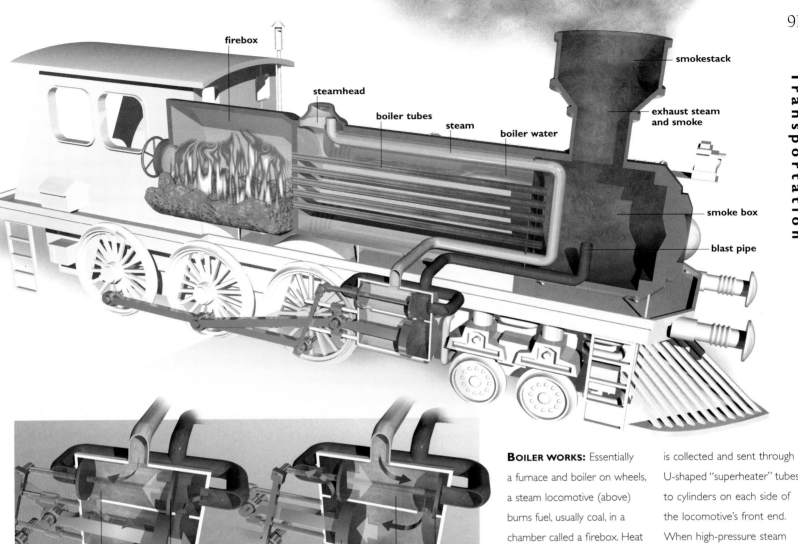

firebox

steamhead

boiler tubes

steam

boiler water

smokestack

exhaust steam and smoke

smoke box

blast pipe

cylinders

valve piston

drive piston

drive piston

valve piston

cylinders

BOILER WORKS: Essentially a furnace and boiler on wheels, a steam locomotive (above) burns fuel, usually coal, in a chamber called a firebox. Heat passes through tubes inside the huge water-filled boiler and generates steam, which is collected and sent through U-shaped "superheater" tubes to cylinders on each side of the locomotive's front end. When high-pressure steam enters the **CYLINDERS** (left), it moves the pistons and drives the train.

GOING THE DISTANCE WITH DIESEL

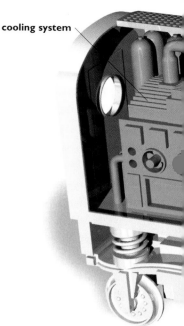

cooling system

SEE ALSO

Riding Rails
• 90

Smart Trains
• 98

THE diesel engine spelled the end of the line for the steam locomotive. Put on the market in 1898 by inventor Rudolf Diesel, a German mechanical engineer, it relies on the same piston strokes and basic moving parts as the gasoline engine. Its fuel, though, is a heavier, thicker, less expensive oil, and it does not need spark plugs for ignition. Instead, this brawny brother of the lighter gasoline engine compresses air inside a cylinder to temperatures that may reach 1000°F and, at the moment of maximum compression, ignites a charge of fuel oil sprayed into the heated air, producing the power stroke.

Because a diesel engine performs more work per gallon of fuel, it is a boon for long-hauling trucks, buses, ships, and trains, and for heavy agricultural and road-building equipment. It works one way in trucks and another in trains, however. In a road vehicle the engine's power is transmitted directly to the wheels, but in a train the engine is connected to an electrical generator. The current produced is stored in huge batteries, then fed to electric motors installed in so-called bogies, which are pivoting carriages that not only house the motors for the driving wheels but also enable the train to negotiate curves. A diesel locomotive is an enormously heavy machine that would be difficult to control if power went straight to the wheels. An electric motor's output can be regulated fairly easily and produces power at very low speeds.

The diesel engine may not evoke images of sleek sports cars and racing boats, but it nonetheless reminds us that function, especially when efficient, is as admirable a quality as form.

RAILROAD YARD

Freight cars pack a railroad yard in Los Angeles, California, just one stop on the United States' 121,000-mile-long rail network, the largest in the world. Following the country's switch from steam to diesel-electric engines, U.S. locomotives became more powerful, less polluting, and cheaper to operate and service. Some 20,000 of them are used solely for hauling freight. Today, rails carry 35 percent of the nation's freight, and trucks carry 39 percent.

LOCOMOTIVE (above): Diesel fuel and electricity combine to provide power without the huff and puff of steam and without overhead wires. The engine connects to a generator making electricity for storage in large batteries alongside the wheels. Power for the driving wheels comes from electric motors that draw on the batteries.

FIRING UP A DIESEL ENGINE (opposite, middle): The piston in the cylinder squeezes and heats air trapped inside; at the top of the stroke, the system injects oil; the air and oil mixture burns and drives the piston down to turn the crankshaft connected to the generator.

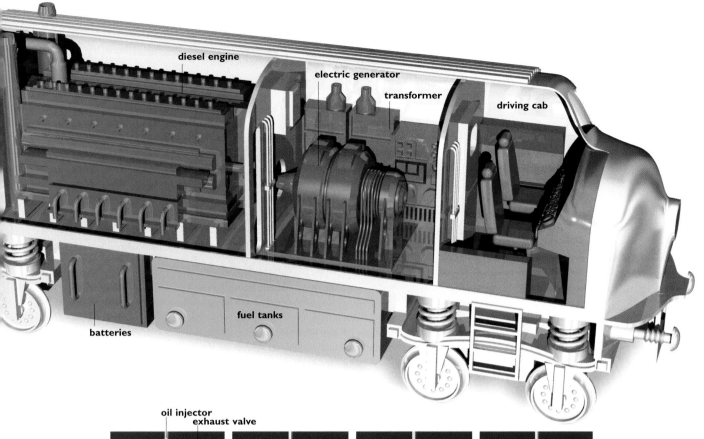

diesel engine

electric generator

transformer

driving cab

batteries

fuel tanks

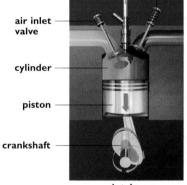

oil injector

exhaust valve

air inlet valve

cylinder

piston

crankshaft

intake

compression

combustion

exhaust

Tons of fuel (below) move in tankers of the Canadian Pacific Railroad. Although heavier and thicker than gasoline, diesel oil—a burner with attitude—performs efficiently and for less money.

SEE ALSO

**Fast
Track • 96**

**Making
Electricity
• 54**

**Smart
Trains • 98**

IN 1873, Andrew Smith Hallidie, an American engineer and inventor, patented the first cable cars: rickety trolleys hauled up San Francisco's steep hills by an endless wire-rope cable running in a slot between the rails. Drawn by a steam-driven mechanism in a powerhouse, the cable cars—still operating today but with technical improvements—eventually spelled the end of horse power as a means of moving passenger coaches. Other cities, including Los Angeles, Washington, D.C., and Kansas City, followed suit, but traffic congestion was a major problem. As a popular joke in late 19th-century Boston had it, the city's streets were so clogged by trolleys that passengers could reach their destination more quickly by climbing onto their car's roof and walking across the tops of stalled vehicles.

The electrified subway changed all that. The world's first subway, run with steam locomotives, opened in England in 1863 and converted to electricity in 1893. That year, Budapest, Hungary, became the first European city to build an electric subway line. America's first electric subway system opened in Boston in 1897, followed in 1904 by the one in New York City.

Subway trains run on a pair of rails in systems that may include hundreds of miles of track, and they are generally powered by voltage running through a third rail. Electric surface trains work on the same principle, but they usually pick up voltage from overhead lines with a folding, scaffold-like apparatus known as a pantograph. The same arrangement drives electric buses, sometimes known as trackless trolleys. Monorails, which are elevated trains built for relatively short distances, run on a single rail, either straddling it or hanging beneath it.

Streetcars, such as this one passing the State Opera in Vienna, Austria (above), draw electric current from overhead wires via a spring-loaded pantograph on the roof.

In San Francisco (opposite), a celebrated cable car gives its passengers a white-knuckle ride on California Street.

Metro blur (below): A subway train flashes from one station to the next in Washington, D. C., carrying commuters and tourists. The capital city's extensive public transportation system has alleviated, to some extent, the surface traffic congestion in the region.

New York's Third Avenue "L" (right)—elevated above the city street—makes a rattling run past the rooftops of Gotham in 1916.

ON THE FAST TRACK

SEE ALSO

On, Over, Under Streets • 94

Riding Rails • 90

Skimming Waves • 104

TOP speed for a subway train is around 75 miles an hour, and for the fastest U.S. train—Amtrak's Metroliner—it's about 125 mph. The top speed for most of the early steam locomotives was only about 25 mph, but it should be noted that in 1893 a steam engine known as the 999 thundered between Batavia and Buffalo, in New York, at 112.3 mph, the fastest humans had ever moved.

Today, France's electric streaker, the TGV (Train à Grande Vitesse) hits 185 mph and is matched or even surpassed by Japan's recent 500-series Nozomi trains. The era of magnetic levitation (maglev) trains, likely to arrive in the 21st century, promises more speed. Magnets on the underside of a maglev and in the track lift the train and propel it with shifting magnetic fields. The idea is not new. In the 1960s, James Powell of Brookhaven National Laboratory was stuck in a New York traffic jam, wishing he could fly over it. A few years later, he and a colleague patented a magnetic levitation system. Eventually, MIT's Henry Kolm built a scale model of a contraption he called the magneplane, and though it moved only a fraction of an inch above a 400-foot aluminum track, it reached an astonishing 56 mph. Japanese scientists, among others, have tested maglev models; in 1979 they got one to run at 321 mph.

MAGNETIC POWER (below): A maglev train puts the principles of attraction and repulsion to work. Electromagnets on the train's underside (shown in red) and on the track (yellow) float the train on a magnetic cushion with no ground contact. Fed with alternating current, propulsion magnets in the sides of a U-shaped aluminum guideway pull and push the train along and control its speed in a smooth sequence of events. Swift and comfortable, maglev trains will give new meaning to "ground transportation."

Testing a prototype (right): On Kyushu, Japan's southern island, a 72-foot, 17-ton test maglev hit speeds in excess of 300 miles an hour along a 4-mile track in Miyazake Prefecture.

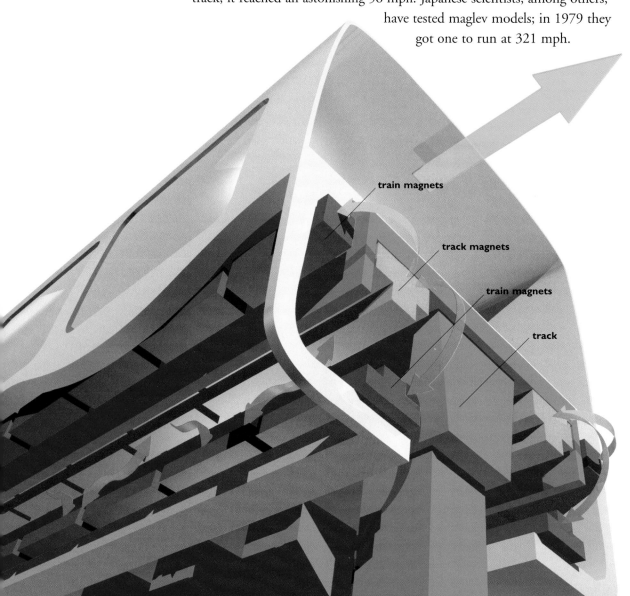

train magnets

track magnets

train magnets

track

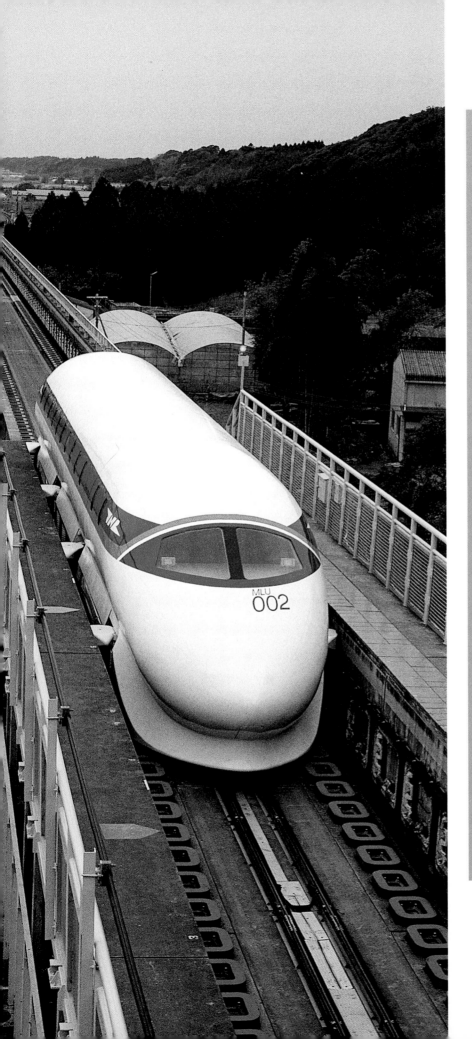

SUPERCONDUCTOR

When a metal becomes superconducting—that is, it transports electricity indefinitely without loss from resistance when subjected to intense cold—it produces a magnetic field with almost magical properties. Here, a thallium-based superconductor bathed in supercold nitrogen vapor floats magnets above and below it through repellent magnetic force. The potential is great because superconductors have the ability to affect everything that uses electricity. Their impact in the near future will be strongest in the fields of computer and information technology and in maglev transportation.

SMART TRAINS, SMART TRACKS

SEE ALSO

Computers • 236

Diesel • 92

On, Over, Under Streets • 94

Riding Rails • 90

Sending Signals • 150

WHATEVER the train—maglev, diesel, or electric—it would be ill-fated in these days of high speed and passenger volume if there were no signals and no control and information systems. Passengers sitting comfortably or standing in railroad and subway cars are blissfully unaware that the safety of their trip, as well as the efficiency of the train, is being monitored by banks of computers in the engineer's cab and in bunker-like, central-control buildings.

A century ago, signalmen positioned along rail routes controlled travel by telegraphing information to one another about their sections of track. Modern train control still depends primarily on dividing the tracks into zones, or blocks, but each is now watched by computers that are fed information by an intricate network of transmitter-receivers, called transponders, on and under the tracks. This trail of electric circuitry activates signals and switching devices, sets speed limits, warns operators of trains in their path, and ensures a safe distance between trains. If a train operator ignores restrictive signals or is incapacitated, the train can automatically be slowed to a permissible speed or braked to a full stop. At the heart of the rail system, in a computer-packed control room holding large, lighted maps, a train's position and speed can be charted in real time by dispatchers, and track switches can be changed by a few strokes on a central-control keyboard.

Traffic control (above): Casey Jones, with a hand on the throttle and his eyes on the tracks, could not have imagined today's train traffic-control centers. A train-spotter at a Union Pacific facility in Nebraska keeps computerized watch over the location and destination of his rolling stock.

Guiding lights (left): Myriad hidden and complex circuits electronically govern modern train signaling, but engineers

must still look for the familiar glow of track lights such as these in Los Angeles, California.

Awake at the switch (right): A signalman in 1925 works on the main line at Charing Cross Station, London. In years past, block-signal systems manually operated at stations along the rail route kept trains safely apart. To get trains from one track to another, a job also performed by hand, workers used a switch stand and lever.

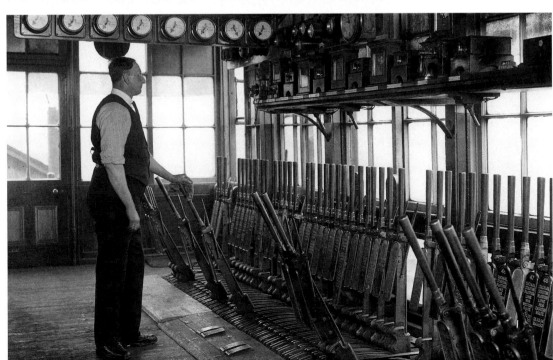

SEE ALSO

Boats Afloat
• 102

**Skimming
Waves • 104**

**Wild Blue
Yonder • 108**

Physics 101 (below): The circular motion of a simple hand crank winds the line that hauls the sail that, in turn, catches the wind that moves a sailboat.

SAILING (below) before the wind is an easy push forward; sailing forward against the wind, or with wind blowing across, requires complex maneuvers, such as setting a zigzag course and changing the sail's angle.

WITHOUT the wind, becalmed sailboats are nearly as still as seascapes hanging preserved in time on museum walls. Only paddles, a motor, or the wind can get things going again. Long ago, primitive peoples moved their vessels across the water with crude paddles, a locomotive power carried to its finest extreme in sleek oar-propelled war galleys that were built in 16th-century Venice.

It is the sail, though, that conjures up images of graceful movement, of nature's own technology at work. "Of all man-made things," wrote a sailor-author, "there is nothing so lovely as a sailboat. It is a living thing with a soul and feelings, responsive as a saddle-horse, loyal as a dog, and thoroughly downright decent."

Whether a sailing vessel is a simple day-sailer or an elaborate, full-rigged merchantman or man-o'-war carrying acres of canvas, it is a wind-driven work of art, and because its sails act as airfoils, it is a nautical version of an airplane. Indeed, although the medium through which sailboats move is water, similar forces of lift, drag, and thrust are at work. Sails are cut and sewn to form a curve much like that of a plane's wings, and lift, or drive, results when wind exerts less pressure on the convex side of a sail. A boat sailing into the wind is pulled diagonally forward by lift generated as wind flows over the sails; running before the wind, it is shoved forward by wind pressure coming from behind.

The complex array of sails and rigging used by the classic square-riggers—20 or more enormous sails in a three-masted ship—provides more drive for the mainsails by compressing, funneling, and deflecting air. The auxiliary sails may also furnish drive themselves.

Yard work (above): Crew members aboard the square-rigged *Sea Cloud* toil amid a cat's cradle of rigging; the sailing ship's numerous lines work specific sails and support masts and spars.

Crow's-eye view (opposite) from a contender in the international Whitbread Race reveals sails trimmed at a 45-degree angle. This position catches wind coming from the left and deflects it astern, driving the boat forward.

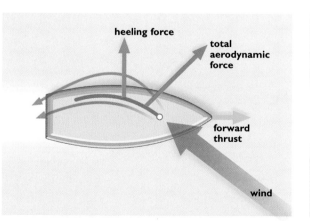

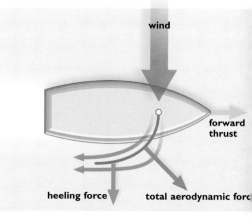

Boats Afloat

See also

Hoisting Sail
• 100

Skimming Waves • 104

Tin Fish • 106

ACCORDING to legend, the Greek physicist and mathematician Archimedes (287-212 B.C.) was sitting in his bathtub, taking note of the amount of water his body had caused to overflow onto the floor, when he formulated the basic principle of buoyancy. Thus began the science of hydrostatics, which deals with the laws of fluids at rest and under pressure. Its main point is that a body immersed in fluid loses weight equal to the weight of the fluid displaced. Put another way, things that float—ships and swimmers, ducks and dugouts—do so when the weight of water they displace is exactly the same as their own. If, however, an object displaces a weight of water less than its own, it sinks.

But there's more to it than weight. A wooden or fiberglass platform will not float if too many heavy rocks are put on it, and neither will a bracelet or chunk of steel placed on the water's surface. They sink because their density is greater than that of water and because they do not displace enough water to create the upward force necessary for flotation. On the other hand, wooden boats and steel ships float because their hollow configuration makes their average density less than that of water. Their weight is spread over a larger volume and they displace more water than, say, a steel block. We float, too, when we inhale and rest on our backs in a lake, and we sink when we exhale and point our feet downward.

Virtual ship (below): In Hamburg, Germany, technicians make a dry run on the bridge of a ship simulator. Such non-sailing "vessels" mimic a ship's motions in various marine environment conditions.

The bulbous bow (opposite, bottom left) of a ship under construction at the Ulsan Hyundai shipyard in South Korea will improve stability, reduce drag, and increase the vessel's buoyancy.

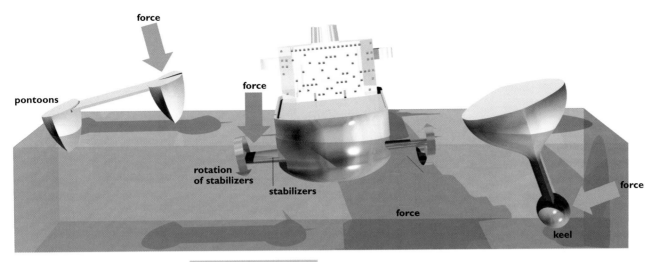

force

pontoons

force

rotation
of stabilizers

stabilizers

force

force

keel

STABILITY OF A SHIP (above) depends on the vessel's center of gravity. Keels and pontoons help maintain stability and minimize the risk of capsizing, but very large vessels may employ stabilizers, winglike protuberances that provide lift to counteract the sea's roll.

GYROSCOPE

Devised by French physicist Jean Foucault in the 19th century, a gyroscope is a rotating disk mounted so that its axis turns freely in any direction. Because it has the ability to hold the same position in space no matter which direction its base is tilted or turned, a gyroscope can control a ship's stabilizers.

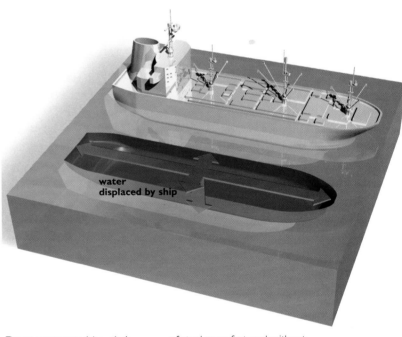

water
displaced by ship

DISPLACEMENT (above): A ship floats because it creates an upthrust force from the water equaling the vessel's own weight, and it does so before the point of submersion. For example: A thick, flat slab of steel manufactured without a hollowed out, bowl-shaped hull to distribute the slab's weight would quickly sink like a stone. Design can also affect speed; the deeper a boat sits in water, the slower it goes.

SKIMMING THE WAVES

SEE ALSO

Boats Afloat
• 102

Fast Track • 96

Mediums and Messages • 214

ALL boats displace water when at rest and create bow waves when they move. Planing boats—powerboats, aquatic scooters, and hydrofoils, for example—climb up over their bow waves as they pick up speed and skim the surface. In the case of hydrofoils, underwater "wings" lift them free of the water as they gain speed, enabling the vessels, driven by turbine-powered pumps and water-jets, to skim along above the surface. So-called hovercraft rise higher off the water because huge fans create cushions of air beneath them.

The most innovative craft may be the magship, a vessel that works on the same principle that drives the maglev train. Conceived in the 1960s by Stewart Way, an American engineer, it had a test run as a 10-foot-long, submarine-like model that ran at 2 knots for about 12 minutes. Years later, Japanese physicist Yoshiro Saji constructed a 12-foot, wooden-hulled model that ran at 1.5 knots.

A magship is propelled by the same force that spins rotors in electric motors: electromagnetism. Under the dictates of a principle known as Fleming's Left-Hand Rule (a magnetic field plus electric current produces a linear force), the ship is an ingenious version of a DC motor, except that it has no moving parts. Instead of wire, the conductor of current is the ocean, a medium full of dissolved salts that are excellent carriers of electricity. An onboard generator sends current into the water between electrodes attached to the hull's underside. At the same time, superconducting magnets on the hull beam a powerful magnetic field into the water. Because the current flows at right angles to the field, force is exerted against the seawater, driving it backward and thrusting the ship forward.

A hydrofoil (below) carries passengers from one island to another in Hawaii. Because much of the vessel's hull does not touch the sea surface and the foils have minimal contact with the waves, the hydrofoil's water resistance drops, and the boat achieves high speeds.

"WATER WINGS" (opposite, bottom) serve as the key element in a hydrofoil's design. Shaped somewhat like the wings of an aircraft, they work in much the same way: Higher pressure beneath them than above them creates lift.

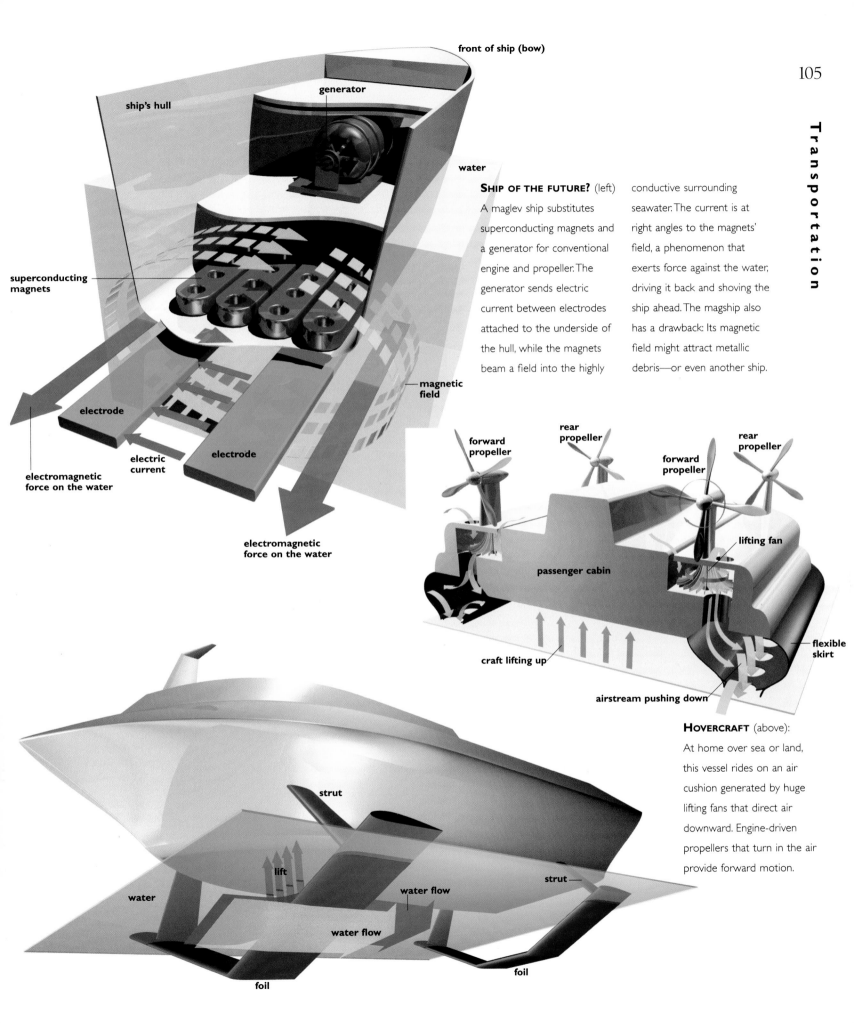

front of ship (bow)

generator

ship's hull

water

SHIP OF THE FUTURE? (left) A maglev ship substitutes superconducting magnets and a generator for conventional engine and propeller. The generator sends electric current between electrodes attached to the underside of the hull, while the magnets beam a field into the highly conductive surrounding seawater. The current is at right angles to the magnets' field, a phenomenon that exerts force against the water, driving it back and shoving the ship ahead. The magship also has a drawback: Its magnetic field might attract metallic debris—or even another ship.

superconducting magnets

magnetic field

electrode

electrode

electrode

electric current

electromagnetic force on the water

electromagnetic force on the water

forward propeller

rear propeller

forward propeller

rear propeller

lifting fan

passenger cabin

flexible skirt

craft lifting up

airstream pushing down

HOVERCRAFT (above): At home over sea or land, this vessel rides on an air cushion generated by huge lifting fans that direct air downward. Engine-driven propellers that turn in the air provide forward motion.

strut

strut

lift

water

water flow

water flow

foil

foil

TIN FISH

SEE ALSO

Boats Afloat
• 102

**Nuclear
Power** • 58

NO ship at sea is more awesome than the nuclear-powered submarine. It is a sleek, gray, cigar-shaped ship loaded with electronics, armed with missiles, mines, and torpedoes, and fueled by a heat-producing reactor.

Submarines dive, rise, and float under the water by adjusting the amount of water and air in their ballast tanks. On the surface, the tanks are full of air, making a ship weigh less than the volume of water it displaces. Flooding the tanks causes a submarine to sink because it then weighs more than the water it displaces. Rising requires a vessel to reduce its weight by forcing compressed air into its tanks to expel seawater. To float just beneath the surface, the amount of water in the tanks must equal the weight of the water displaced.

While the detailed technology behind nuclear propulsion is highly classified information, one can safely say that a submarine travels essentially on steam. Intense heat is produced from the fission of nuclear fuel in the ship's heavily shielded reactor. This heat generates steam, which drives not only the turbine generators that supply electricity to the ship but also the main propulsion turbines that turn the propeller.

Tanks full of air, a nuclear-powered submarine (upper right) lolls atop the ocean.

SUBMARINE DESIGN (lower right): A rounded configuration and double-walled hull resist crushing deepwater ocean pressures. A submarine can dive, travel underwater, rise, and run on the surface—all by adjusting the volume of water and air in ballast tanks. A nuclear reactor generates steam that drives turbines turning the propeller.

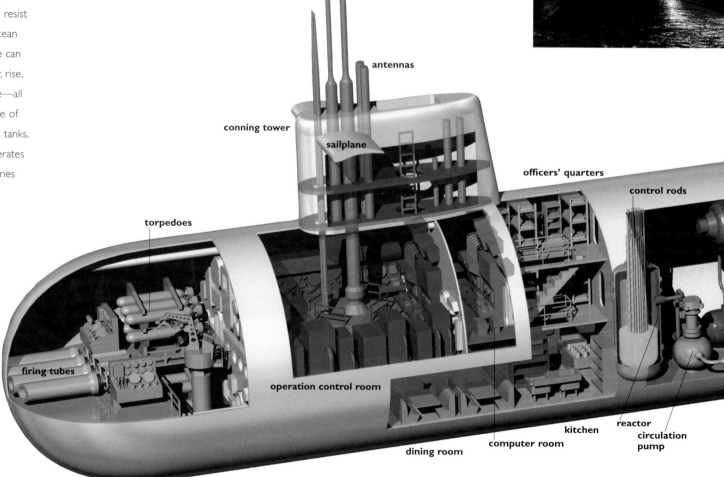

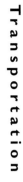

Reflected waves return
to sub in deeper water.

Sound waves from deeper sub
travel out and bounce off sub
in shallower water.

sound waves

steam

water separation tank

propulsion
turbine

condenser

heat exchanger

turbogenerator

water tank

water-feed tank

SONAR (above), which stands for "sound navigation and ranging," describes a device or a method that transmits electronically generated sound waves through water. Computers tuned to the pinging echoes, or reflected sound waves, can determine the direction from which they come and calculate the time they take to return, processing the information and using it to navigate the vessel, measure depth, map the seabed, and detect underwater objects.

Submariners (top left) in the control room of the USS *Billfish* perform highly complex operations. Using electricity supplied by generators and making fresh water and oxygen from the ocean, crews can live submerged for months.

INTO THE WILD BLUE YONDER

SEE ALSO

Boats Afloat
• 102

**Skimming
Waves** • 104

Up and Away
• 112

Wright Stuff
• 110

EVER since the ancient Chinese began to experiment with man-flying kites and parachutes, and perhaps even earlier, people have tried to emulate birds. Indeed, the aerodynamic principles and forces that govern our aircraft, from gliders to jetliners to rocket ships, are aptly demonstrated in the flight of birds. Simply put, they involve lift, drag, thrust, and weight.

Air hitting a wing's leading edge during flight and streaming over the top and bottom surfaces of the airfoils, or wings, provides the lift that raises a craft and keeps it "afloat," no easy task, considering that lift must equal the craft's weight. Enter Bernoulli's principle: Pressure is inversely related to velocity. Put another way, fast-moving air exerts less pressure than slow-moving air. During flight, the airflow over the longer upper surface of a wing travels faster than air on the underside, producing less pressure. The net force on the wing is an upward force exerted by the slower-moving stream of air beneath it. With good wing design, the lifting force may be magnified many times beyond that caused by the impact of air on a fairly basic airfoil.

Lift, however, must fight against drag, a force caused by friction as the plane moves through the air and by changes in airflow. Drag slows the plane, requiring the thrust of an engine to compensate and keep the plane moving.

Supersonic transport:

The *Concorde* (right) rises into the sky. Using forces different from those that keep ships afloat, planes must generate lift greater than their own weight during takeoff—and equal to it to stay aloft. Speed and wing surfaces provide the great lifting force needed for takeoff; along with landing, takeoff ranks as the most crucial of flying maneuvers.

AIRLIFT (lower right): Air that flows over the top of a wing moves faster and exerts less pressure than the stream beneath; the greater pressure of the slower-moving air below the wing lifts the airplane.

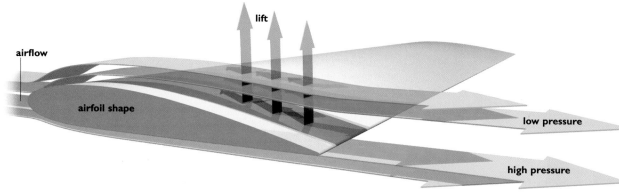

lift

airflow

airfoil shape

low pressure

high pressure

A stunt pilot (left) puts a fast-moving light plane through its many paces, a workout that shows how a single engine and the skill of a lone pilot can use lift and thrust to overcome drag and gravity.

MOVABLE SLATS (lower left) at the front of a wing improve airflow, and flaps at the rear enhance wing curvature and area. Closed in normal flight, they open at takeoff and landing, increasing lift. A plane stays down after landing when other hinged surfaces rise at right angles to the wings, breaking airflow.

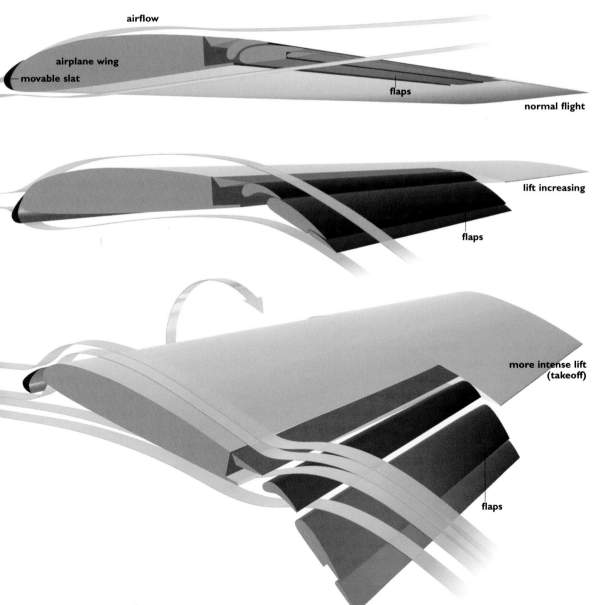

airflow

airplane wing

movable slat

flaps

normal flight

lift increasing

flaps

more intense lift (takeoff)

flaps

BALLOONS

With the grace and effortlessness of a jellyfish suspended in its natural habitat, Steve Fossett's *Challenger* (above) rises into the sky because it is lighter than the air. Kept aloft by helium or by air that expands when heated by butane or propane burners—a vast improvement over brazier-fired and hydrogen-filled bags that once were used—a balloon uses the same forces that keep a ship afloat. Fossett's balloon gets power from a solar panel.

THE WRIGHT STUFF

SEE ALSO

Automobile
• 82

Up and Away
• 112

**Wild Blue
Yonder** • 108

**Workshop
Tools** • 44

THE Wright brothers' pioneering flying machine was powered by a four-cylinder, twelve-horsepower gasoline engine; its carburetor was a tomato can. Airplanes that came along later were able to get by without tomato cans, but they also relied on simple internal combustion engines to drive propellers that converted engine shaft torque, or turning force, into thrust. Propellers have blades that are shaped like wings: The front surface of each blade is more curved than the back, so a forward aerodynamic force is produced.

A jet engine works in much the same way, although it doesn't look like a piston engine. A commonly used example explaining some of the principles behind its operation is adequate only insofar as it goes: If you inflate a balloon and release it untied, it will fizz about a room until the escaping air is depleted. A jet engine is basically an internal combustion engine; however, it uses the energy produced by combustion directly, and it does not need pistons to transmit driving power. Large quantities of air are drawn into the engine, compressed by a bladed turbine, and sprayed with kerosene fuel. When the mixture is ignited and the temperature in the combustion chamber exceeds 2500°F, the heated, expanding gases rush through an exhaust nozzle, providing the tremendous thrust that drives the plane forward.

Gears (left) put teeth into moving machinery. A turbo-prop engine uses this Allison planetary reduction gear.

PROP POWER (below left): Propeller blades resemble wings and depend on the same forces to produce lift.

Bug-hunting (below): A technician checks an engine. Computer-assisted "smart test" equipment now identifies glitches once spotted by eye.

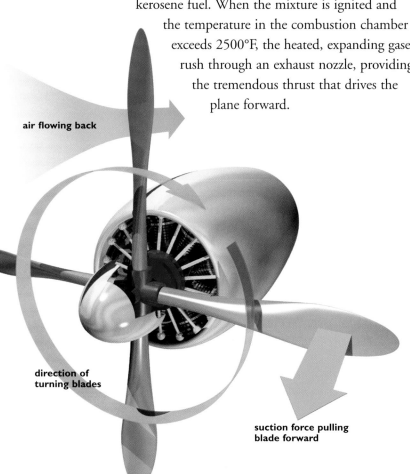

air flowing back

direction of
turning blades

suction force pulling
blade forward

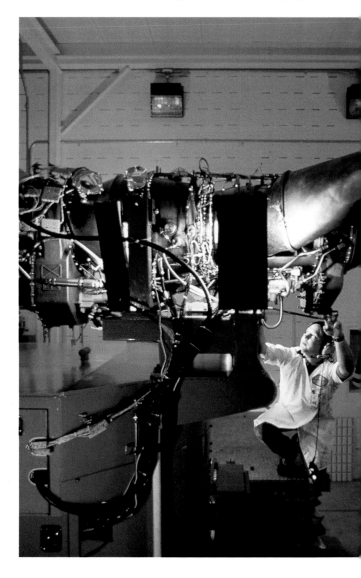

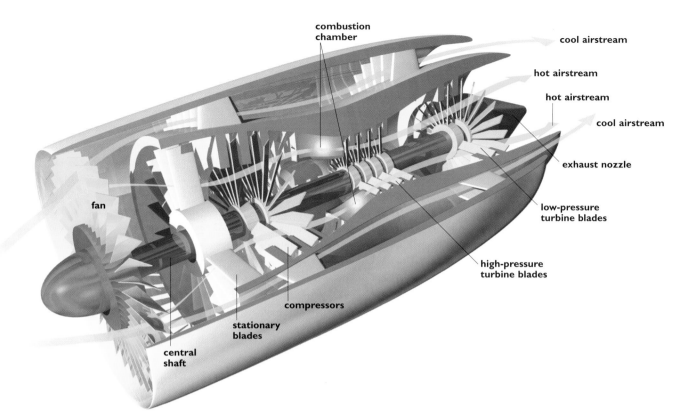

combustion chamber

cool airstream

hot airstream

hot airstream

cool airstream

exhaust nozzle

low-pressure turbine blades

high-pressure turbine blades

fan

central shaft

stationary blades

compressors

PROPULSION SYSTEM

(above): The action-reaction principle drives a turbojet engine. Exhaust gas provides the thrust and simultaneously drives the turbine that turns the air-intake fan. The turbine compresses drawn-in air to improve combustion and make the exhaust work better.

Huge fan-blades (left) inside a jet engine dwarf an engineer. Having fewer working parts than a propeller engine and fed on relatively cheap kerosene, jet engines provide enormous thrust and guzzle fuel. For a transatlantic crossing, a fully loaded jumbo jet may require a fuel supply of more than 40,000 gallons. Each one of its engines can burn as much as 1,000 gallons of fuel per hour.

Up, Up, and Away

See also

Supersonic • 116

Wild Blue Yonder • 108

Wright Stuff • 110

Leonardo da Vinci is credited with designing a flying machine that could twirl itself skyward with the aid of a helix, or screw. He was inspired, perhaps, by "bamboo dragonflies"—the name given by fourth-century Chinese to toplike toys that could climb into the air with the pull of a string. The inventor was unable to build a working model, however, because the engine to power it had not been invented.

The first successful helicopter was invented by Igor Sikorsky, a Russian-born U.S. aircraft designer, and first flown in 1939. But the basic design was pure Leonardo, and the principles behind it were the same ones that govern all powered aircraft. The difference, of course, is in the function of a helicopter's "wings"—actually the rotor blades that enable it to ascend and descend vertically, to hover, and to fly in any direction.

Because a helicopter's moving wings provide both lift and thrust, the angle at which each blade enters the air, called the pitch, is essential to the craft's control. As the blades turn, the pilot uses the collective pitch stick to increase their pitch equally and thus lift the helicopter vertically. When the pilot decreases the pitch, the helicopter descends. To hover, the blades are angled just enough to produce lift equal to the craft's weight.

Propelling a craft forward, backward, or sideways requires a pilot to use a cyclic control stick that tilts each blade at precise moments as it revolves. Such relative ease of operation, however, has a downside: the action-reaction law. A helicopter having only an overhead rotor would be forced in the direction opposite to the rotating blades and would spin out of control. To compensate, a tail rotor shoves air to one side and keeps the craft on the right path.

To track ice (above) in the St. Lawrence River, cockpit observers rely on a helicopter's exceptional maneuverability, including its ability to fly in any horizontal direction without changing its alignment.

Rotary motion (far left): A tail-mounted vertical rotor produces a sideways thrust that prevents a helicopter from spinning out of control. It also helps the pilot steer: Decreasing the thrust turns the craft one way; increasing thrust turns it in the opposite direction. Tilting the main rotor blades in unison keeps a craft on a straight course, and by tilting them in sequence, the pilot increases lift on one side to move the craft left or right.

Touchdown (left): A U.S. Navy minesweeping helicopter comes in for a trademark vertical landing on the deck of the USS *Nashville*, a maneuver that requires minimal space.

sideways thrust from tail rotor

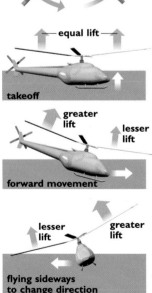

equal lift

takeoff

greater lift

lesser lift

forward movement

lesser lift

greater lift

flying sideways to change direction

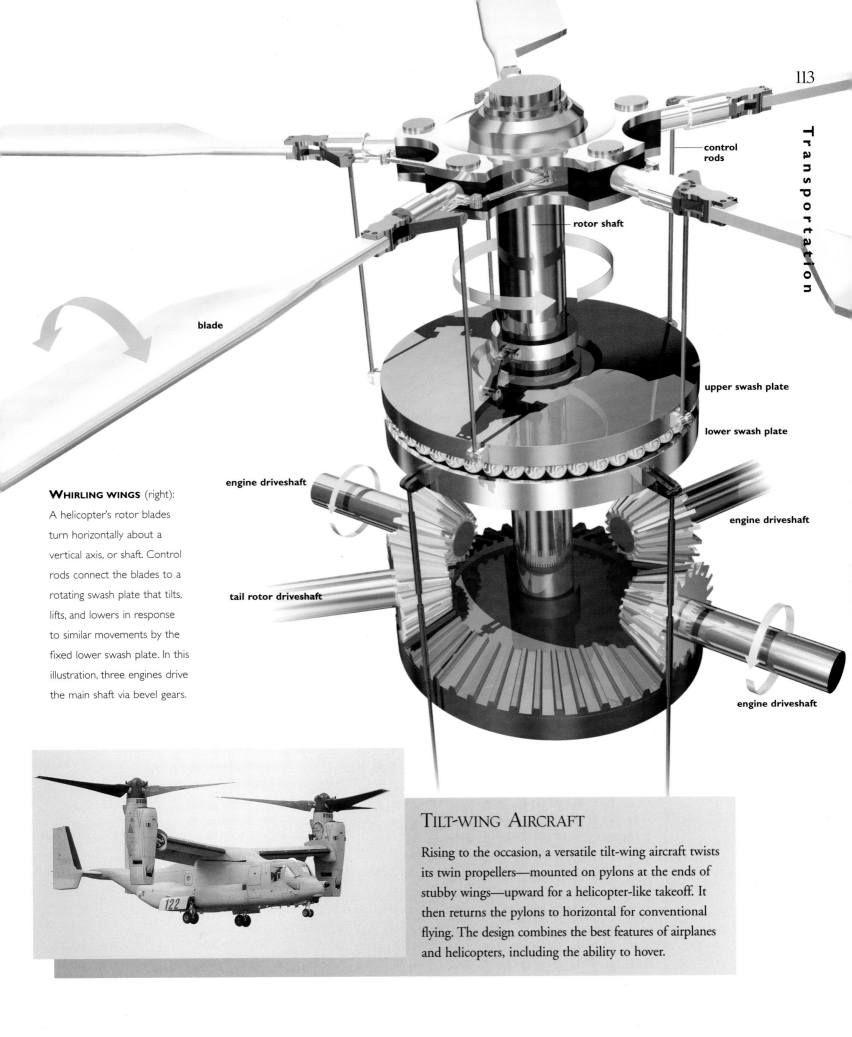

control rods

rotor shaft

blade

upper swash plate

lower swash plate

engine driveshaft

engine driveshaft

tail rotor driveshaft

engine driveshaft

WHIRLING WINGS (right): A helicopter's rotor blades turn horizontally about a vertical axis, or shaft. Control rods connect the blades to a rotating swash plate that tilts, lifts, and lowers in response to similar movements by the fixed lower swash plate. In this illustration, three engines drive the main shaft via bevel gears.

122

TILT-WING AIRCRAFT

Rising to the occasion, a versatile tilt-wing aircraft twists its twin propellers—mounted on pylons at the ends of stubby wings—upward for a helicopter-like takeoff. It then returns the pylons to horizontal for conventional flying. The design combines the best features of airplanes and helicopters, including the ability to hover.

SEE ALSO

Computers • 236

Mediums and Messages • 214

Tin Fish • 106

TODAY, from takeoff to landing, commercial pilots fly by electronics, radio beacons, and radar. From the ground, airplanes are tracked and directed by watchful controllers in towers packed with technology's best eyes and ears. In the air, they are guided and monitored by onboard computers and sensors that control aircraft movements via fiber-optic cables. Pilots use the cables to transmit coded digital signals directly to the motors moving the control surfaces.

Computers and software also establish safety parameters, check speed and direction, and correct for weather vagaries and possible pilot miscues. Weather radar in the nose sends information to the plane's data system; sensors detect wind shear and monitor fuel consumption. The cockpit gauges, which were once a bewildering array of more than a hundred dials, now consist of a relatively few glowing monitors. Under the direction of the avionics system—a blend of navigation, flight management, and digital automatic flight-control systems—only two people are generally required in the cockpit. Indeed, modern aircraft are capable of flying by themselves during and immediately after takeoff.

Landing a Boeing 757-200 (opposite) at night requires both superb piloting skills and knowledge of avionics, the science of aviation electronics. Typically, a jetliner's glowing cockpit includes standbys such as airspeed indicators and altimeters, as well as complex instrument-landing and navigation systems.

ROTATING GROUND RADAR ANTENNAS (below) beam radar signals at an airliner and receive reflected pulses augmented with signals from the radar beacons on the underside of the aircraft. Controllers use these signals to determine the approach, position, altitude, and identity of the plane.

Controllers in a tower (upper right and inset) use images on their radar screens to direct the incoming planes.

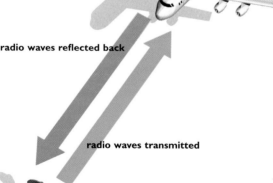

radar antenna

radar antenna

radio waves reflected back

radio waves transmitted

radar antenna

radar display screen monitored by air-traffic controller

control tower

From Supersonic to Hypersonic

See also

Alternatives
• 86

Up and Away
• 112

Video Games
• 166

**Wild Blue
Yonder • 108**

SOUND, as many of us have forgotten, travels at 1,090 feet a second in air, a figure that translates into Mach 1 in an aeronautical engineer's lexicon. The first time an aircraft flew faster than that was in 1947, when Capt. Chuck Yeager, USAF, took the controls of an experimental, rocket-powered Bell X-1. Today, while military fighters routinely break through the sound barrier by exceeding 720 miles an hour, only one commercial airliner flies at supersonic speed: the 100-seat, fuel-guzzling, ear-splitting *Concorde,* which streaks through the skies at Mach 2. Now more than 20 years old, the beak-nosed craft is in danger of becoming obsolete. The *Concorde* has limited passenger space, and the plane's impact on the environment is potentially disastrous.

A new generation of experimental hypersonic airplanes capable of flying 10 times the speed of sound—7,200 miles an hour—is on NASA's and airline manufacturers' drawing boards. Fueled by hydrogen, which requires oxygen to burn, a so-called Hyper-X plane uses "air-breathing" engine technology that eliminates the heavy oxygen tanks rockets must carry; instead, it has a propulsion system that draws oxygen from the atmosphere as the plane speeds along. Nearer to home, airliners that now hold some 500 passengers will eventually carry 1,000. Constructed of lighter materials, they will have far more range and will fly at supersonic speeds (at the moment that concept may not be economically wise). NASA's X-38 Crew Return Vehicle, designed to travel to and from orbit, will become the emergency lifeboat for occupants of the International Space Station beginning in 2003.

Speed isn't all that the future of aviation holds. Solar-powered aircraft have already been tested successfully, their propellers driven by thin energy panels coated with solar cells. Innovative designs are also key. Swept-back wings that reduce air shock at supersonic speeds will be common. Fiberglass and metal alloys will allow lighter, fuel-saving construction. Many planes will be more versatile: Jump jets already in use can take off vertically by swiveling jet nozzles downward; when airborne, they fly forward by turning the same jets backward. A tilt-wing aircraft twists its wings and propellers upward for a helicopter-like takeoff and then drops them to the horizontal position for conventional flying. Cockpits of the future will rely on virtual reality: In commercial airliners, computer-generated displays of travel routes and obstacles will surround pilots, replacing windshields. Fighter planes may become fully pressurized, letting pilots work without breathing equipment while projecting essential flight imagery overhead across the canopy.

***Stealth* bomber** (above): Streaking through the sky, this plane looks like a work of abstract art. Its unique shape, sophisticated electronics, and special surface materials give it the ability to evade radar and infrared monitoring.

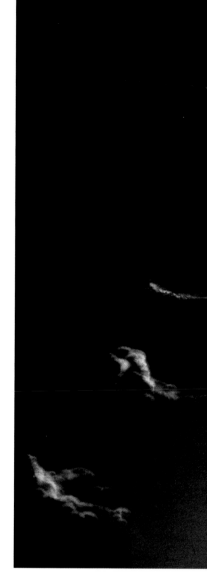

Aerial pogo (below): An experimental, jet-powered, two-man taxi blasts off near Niagara Falls, New York, on Independence Day, 1967.

High in the atmosphere (above), a German-designed Horus spacecraft leaves its carrier after a conventional runway takeoff and streaks for orbit on its own rockets. No longer science fiction, such rocket ships will ply space early in the new millennium.

Flying saucer (right): Hovering over Fort Benning, Georgia, a robotic helicopter can take off, fly, and land by itself, its destination charted by a "pilot" using only a mouse and a computer screen.

ING

FORMAL plant cultivation and animal domestication began around 8000 to 6000 B.C., dramatically changing what humans ate, wore, and used for shelter. Once people discovered that some plants carried the germs of food in their seeds, planting may have been as simple as dropping a few pods into the ground. Or the planting may have already been done by chance, perhaps with seeds fallen from the clothes and wares of nomads, or shaken from the hair of their animals. Today, agriculture is scientific: It is a world that relies on "pre-emerge" herbicides, "post-emerge" chemical treatments, crop rotation, and contour plowing. It produces hybrids and protein-enhanced soybeans. It uses sensors to monitor soil properties and satellites to map planting sites. It opens and cleans cultured shellfish with specialized machinery and sends fleets of computer-guided, automated farm vehicles across fields of grain.

A field-mowing combine harvester spews out the cereal grain for a nation's breakfast tables.

SMART FARMING

SEE ALSO

Computers
• 236

Eyes and Ears
• 220

Sweeter, Leaner, More Tolerant
• 124

Q & A (above): How much moisture is in the soil? How much runoff? What's the humidity? A farmer gets the answers with his field call box, an electronic irrigation and weather-sensor system.

You say tomato, and it says maybe. An electronic color sorter (right) scans the fleshy fruit of a nightshade plant.

From space to soil: Mapped by a Global Positioning System satellite and converted to a grid, a planting field shows up on a computer screen (far right). Such imagery helps today's farmers monitor soil types, watch for disease and insect problems, and control the application of fertilizer to exact field locations, thereby improving crop yields.

NO longer can it be said that the cellular and environmental events that make for inferior crops and low yields are natural, irreversible acts. Science and technology have descended on farms like savior-angels.

The phrase currently in vogue is "precision agriculture." In it, satellite imaging of field conditions helps to improve crops. The light beams of sensors analyze the ability of seed furrows to accommodate the flow of air and water. Tractor-mounted computers and satellite connections identify variations in nitrogen and pH levels, anticipating fertilizer and pesticide needs within feet of a tractor's position. "Precision agriculture helps producers apply the exact amount of chemicals they need, where it's needed, when it's needed," Deputy Secretary Richard Rominger of the U.S. Department of Agriculture has observed. "By the condition, not the calendar. By the foot, not the field. With precision agriculture, every farm, every field, every spot in the field becomes an experimental site." At this writing, two dozen satellites are orbiting the Earth in six paths, mapping out fields in a mosaic pattern.

Agriculture as a whole has taken a cue from such precision. Farmers these days are just as likely to rely on weather balloons carrying instruments that measure temperature and humidity as they are to gaze at thermometers or barometers. Other farmers stroll their fields, monitoring vegetation and soil properties through sensors that are attached to yokes strapped to their backs; microwave radiometers mounted on trucks quickly measure soil moisture.

Farming methods also are more conservation-oriented today. In no-till farming, for example, all or part of the current crop residue is left lying on the soil surface after harvesting has been done. This residue provides a protective blanket and a bed of organic material that conserves moisture. Later, farmers make small grooves in it and drop in seeds. To control the growth of weeds, they rely on herbicide applications rather than tillage.

SEE ALSO

Cutting Edges
• 46

Grass Cutters
• 48

WITHOUT doubt, Cyrus McCormick's mechanical reaper of 1831—a horse-drawn, wheeled contrivance that efficiently gathered and cut bunches of grain stalks—was the most important advance in the mechanization of farming. What has followed is equally impressive: Twine-binders attached to reapers automatically tie stacks of wheat; wheeled seed boxes with drills prepare furrows for planting; weed-blasters use hot air to destroy weeds while sparing the plants; combine harvesters cut fruiting heads, thresh, and clean grain; corn pickers pick, husk, and send corn to shellers that remove the kernels; and cotton pickers twist cotton fiber right off the bolls—a blessing for traditional stoop-laborers. There also are potato-diggers, beet-cultivators, and trunk- or foliage-shakers that drop nuts and cherries into catch-frames.

One new mechanical foliage shaker resembles a giant hairbrush, with 12-foot-long nylon "bristles" that rotate as well as shake. Pulled between rows of orange trees by a tractor, the spikes dig 5 feet into the tree's canopy and shake fruit onto a conveyor belt that transfers it to a self-propelled, bulk-transport vehicle. Simple as it sounds, the technology is expected to dramatically change the U.S. citrus industry. This harvester handles 300 to 400 field boxes of fruit from each acre 15 times faster than hand laborers, and it can harvest a 90-pound field box for 50 cents—a dollar less than the current cost.

"Up from the meadows rich with corn," wrote John Greenleaf Whittier, whose words well describe the golden kernels gushing from this harvester (above right).

A COMBINE (right) does just what its name implies: Moving through a field, it simultaneously reaps, threshes, and cleans grain. The tined reel sets up the grain heads, and a cutter bar slices them. An elevator conveys them to the threshing cylinder, which separates the grain and drops it onto a vibrating pan. The grain then goes to a sieve, and an auger and elevator carry it to the holding tank. The rear beater sends the threshed stalks to straw walkers that eject them for baling.

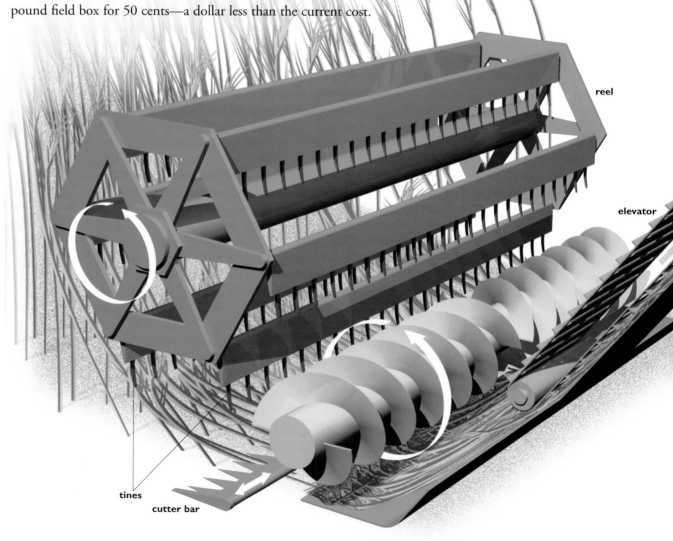

wheat

reel

elevator

tines

cutter bar

Harvesting time (right): A major improvement over the sickles and scythes that reaped grain for centuries, a fleet of harvesters cuts through a Montana wheat field—an unfamiliar sight in 1910, when U.S. farms relied on 24.2 million horses and mules, and only about 1,000 tractors.

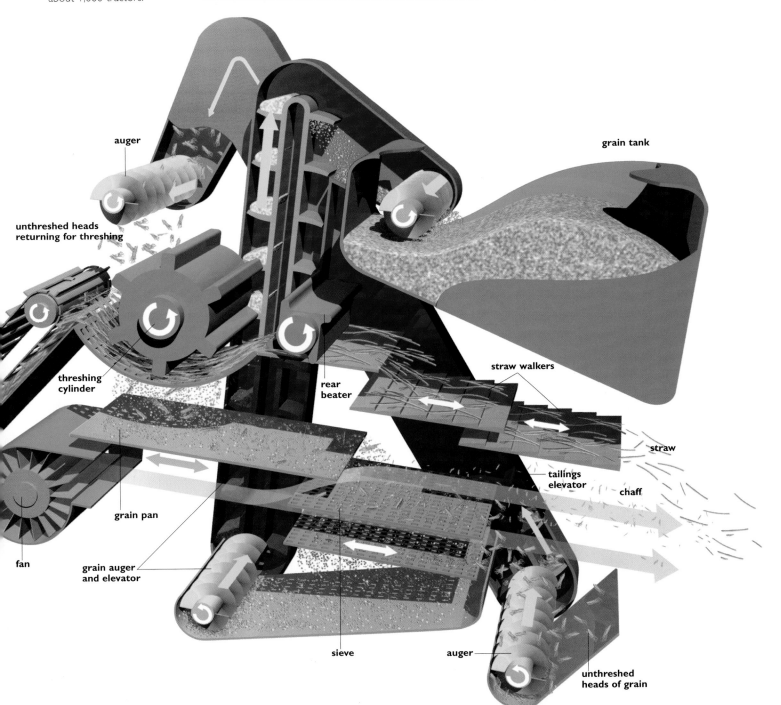

auger

grain tank

unthreshed heads returning for threshing

threshing cylinder

rear beater

straw walkers

straw

tailings elevator

chaff

grain pan

fan

grain auger and elevator

sieve

auger

unthreshed heads of grain

SWEETER, LEANER, MORE TOLERANT

SEE ALSO

Fish Farms
• 126

Smart Farming
• 120

IT tastes like cauliflower when eaten raw, but it is milder, sweeter, and more like broccoli when cooked. Say hello to the broccoflower, a chartreuse cross between the two members of the family Cruciferae. As every health-conscious American knows, these plants are supposed to be very good for us. First marketed ten years ago, the broccoflower—along with FLAVR SAVR tomatoes and SuperSweet onions—is a product of food biotechnology, a broad-ranging science employing a host of techniques to improve the things that we eat. Scientists can genetically alter crops by giving them new, useful genes, thereby shielding them from disease, spoilage, and insects. The plants can then be made to grow with less dependence on chemicals. Crops also can be goaded into faster ripening and higher yields, or they can be endowed with better nutrients and more flavor.

One important example of biotechnology is selective breeding, an ages-old agricultural standby. Essentially, it involves selecting plants and animals with desirable traits and breeding them under controlled conditions, a process that has achieved much success and produced enormous yields. Selective breeding, which concentrates on entire organisms with complete sets of genes, and genetic engineering, which focuses on a few gene transfers, have given us both products and promise. We now have seedless bananas, beardless mussels, leaner animals, and virus-resistant squash. Synthetic Bovine Growth Hormone (BGH), a genetically engineered drug given to cows, increases milk output by augmenting a cow's own natural production of BGH. More than half the cheese produced in the United States is made with chymosin, a biotech-preparation that does away with the need for rennin extracted from calves' stomachs. Tomatoes that can be made to ripen on the vine and then can be shipped without spoiling are a definite plus when you consider that commercial tomatoes are generally shipped green and then ripened with ethylene gas.

Biotech twins (below): Clones Megan and Morag help lead the way to a new world of agriculture, in which farms mass-produce animals bred selectively to give us better food, clothing, and pharmaceuticals. Genetically altered **piglets** (far right)—easy to raise and relatively disease-free—may fill more roles than ordinary pigs. Reprogramming pig genes may result in leaner, more tender meat. Some scientists even talk about inserting human genes, which could turn the animals into factories that churn out organs for human transplant, as well as cells and tissues.

AQUACULTURE: FISH FARMS

SEE ALSO

Hydroponics
• 128

Smart Farming
• 120

Sweeter, Leaner, More Tolerant
• 124

NOT all of our seafood is wild harvest—a fact that might come as a surprise to many people. Aquaculture, the culturing of fish and plants in a controlled area of water, is a rapidly growing agribusiness, larger even than veal, lamb, and mutton combined. According to the U.S. Department of Agriculture, 20 percent of the fish and seafood consumed in the United States is now raised on farms, with catfish, tilapia, salmon, trout, crawfish, and shrimp among the leaders.

Depending on what form of fish or seafood is raised, fish farms may be earthen ponds, tanks, trough-like raceways, net pens, suspended cages, or bottom nets. The "seeds" are fingerlings, very young mollusks, or eggs, which are raised on commercial feed formulated to provide adequate nutrients, or on natural food organisms grown through water fertilizing techniques. Because potentially polluting and disease-bearing waste accumulates in any fish farm, it must be disposed of through elaborate water-circulation systems. Some filtration systems harness bacteria that convert ammonia, which fish secrete through their gills, into nitrates that can be flushed from tanks. As an additional precaution, stocked fish may be treated with antibiotics and other chemicals, a practice that invariably raises concerns among environmentalists and consumers.

One state-of-the-art technique—gene-splicing—can produce catfish, salmon, and trout that not only have more resistance to disease but also are faster growing and more immune to freezing in winter. As with all technology-enhanced food products, concerns and complaints have been raised about them.

Harvesting the sea (below): A Dutch fishing boat hauls its nets out of the sea. In many parts of the world, commercial fishing interests have long relied on draggers and factory ships, which can make huge harvests that deplete stocks of traditional species. While fish farms may not provide as large a nursery as the open sea, they allow better control over feeding and harvesting. Some have great success: A million gallons of water in a tank can yield a million pounds of fish.

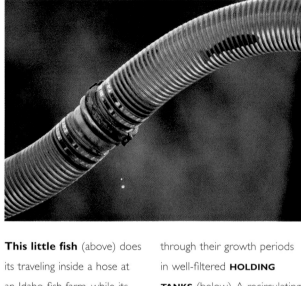

This little fish (above) does its traveling inside a hose at an Idaho fish farm, while its relatives in the wild may ride a tide onto shore or travel upstream against a strong current. Unlike conventional farms, which generally consist of vast tilled fields arranged in familiar patchwork-quilt patterns, water-filled fish farms prove the ultimate in container-growing. In one model, the fish eggs fertilized in an **INCUBATOR BATH** (upper right) eventually go through their growth periods in well-filtered **HOLDING TANKS** (below). A recirculating system cleans out fish waste, removes uneaten food, and controls the water's oxygen levels. Fish farms, which release wastes into natural bodies of water, must also deal with the same problem that faces hog and poultry farms: how to dispose of polluting waste.

eggs and milt from female and male fish

fertilized eggs placed in incubator

larval fish

fry tank

water flowing to fish tanks via a channel

for fish up to one year

for fish from one to two years

for fish two to three years old

eel pass

outlet returning water to river

dam diverting river water

fish ladder

HYDROPONICS: SOILLESS FARMING

SEE ALSO

Fish Farms
• 126

Smart Farming
• 120

CONSIDERING the vulnerability of soil to diseases, pests, changing weather conditions, and inadequate nutrient supply, it is no wonder that farmers long for better control of their plants and sometimes even wish they could control the seasons. One remedy is the science of hydroponics: Plants are grown without soil in nutrient-enriched solutions, with their roots anchored in porous, nonsoil materials. Widely used in botanical research, hydroponics also grows vegetables, fruits, flowers, and herbs in greenhouses and in arid regions of the world.

Soilless culture begins with water enriched by the same balance of nutrient salts found in soil; when dissolved by watering, these nutrients are absorbed by plant roots. The roots themselves are supported by all manner of materials that retain air and water, including sand, gravel, glass wool, rock wool, fiber, and stone. A variation of this process is aeroponics, by which plant roots are suspended in a chamber or a bag; humid air provides the proper environment while a spray mist of nutrient solution keeps the roots moist and nourished. With such midair feeding, almost no water is lost through evaporation, and roots absorb much vital oxygen, increasing metabolism and the rate of growth as much as ten times over that in soil.

The National Aeronautics and Space Administration's interest in growing food in space has led to the development of a unique medium made from zeolite, a common mineral. Zeolite "soil" takes advantage of the mineral's natural properties as a "molecular sieve," allowing it to store and time-release nutrients. Mixing an additive with specially prepared zeolite creates a substance that, in laboratory tests, produces conditions almost comparable to those of the soilless "soil" in conventional hydroponics.

Cool cucumbers:

Pampered by gardeners in the great indoors, cucumbers grow well in a hydroponic garden (opposite, top). Water gardens of all kinds thrived along Africa's Nile River thousands of years ago, and during the Second World War, the U.S. Army grew vegetables hydroponically on infertile Pacific islands. But even though crop yields can exceed the success rate of dirt farming, large-scale soilless farming remains confined to out-of-season greenhouse plants and to areas that have limited arable land.

SLUDGE-BUSTERS

Because the nutrient solutions in the enriched, pure water that makes hydroponics possible are contained, they do not pose the same threat as runoff from fertilized soil. Tainted water pollutes aquifers and presents an immense cleanup challenge, but this is a problem that has a truly natural solution. At an experimental greenhouse (left) in Providence, Rhode Island, sewage is piped into vats, along with a horde of plants, bacteria, snails, and fish that have an appetite for filth. In a few days, the cloudy sample of water filling the flask in the man's left hand will be as clear as the sample in the other flask. It will be nearly drinkable, virtually odorless, and ready for discharge. Bacteria have also been harnessed by scientists who used them to devour environmental pollutants, such as oil spills in the ocean. The bacteria break down the oil in much the same way that ordinary bacteria cause dead animals and plants to decompose and change into soil-enriching substances.

A HYDROPONIC GARDEN

artificially provides all of the ingredients essential for plant growth: oxygen, light, heat, water, nutrients, and carbon dioxide. Inside the protective confines of a greenhouse, plant roots absorb nutrient salts from enriched water. An inert and soil-free porous medium and plastic mesh anchor the roots in the water. Sunlight or artificial "growth" lights ensure that the plants will grow rapidly and healthily. By delivering nutrients directly to the roots, a hydroponic system ensures that energy ordinarily used to produce long roots goes directly into growing a larger plant.

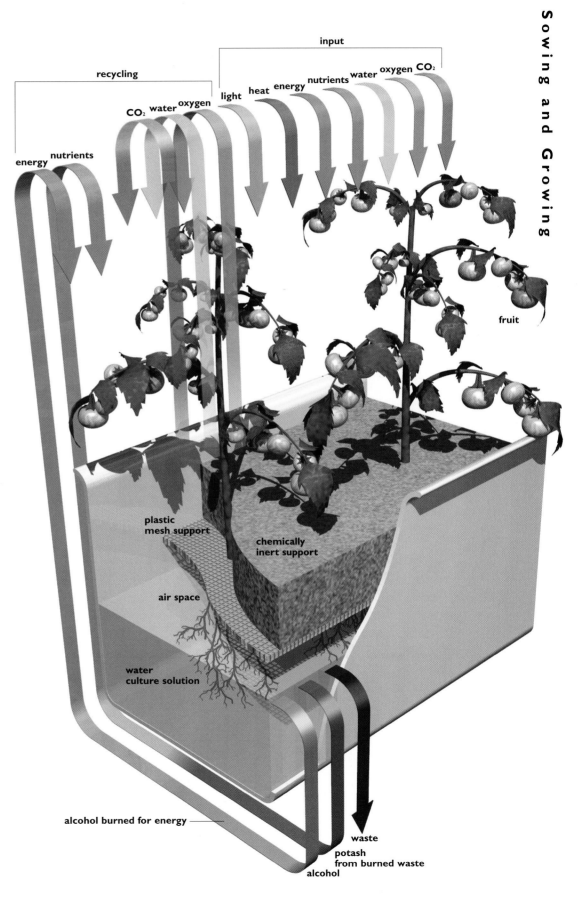

SPINNING and weaving have been around since prehistoric times, when people began using flax and cotton, wool, hemp, and silk to make simple cloth. By the 14th century, Europeans were creating clothing and decorative fabrics from linen, silk, and wool, using virtually all of the basic weaves of today. By the end of the industrial revolution, manual processing had given way to the modernization of techniques and equipment. In today's world, the textile industry is a highly technical enterprise that integrates imaginative design with performance and function. While fibers continue to be made from plant and animal materials, they also are spun from glass and fabricated from organic polymers, which are derived from coal and petroleum. Textiles are among technology's most aesthetic and diverse achievements, and they are key elements in the economies of many nations.

Fine pleats in sheer silk characterized the early 20th-century creations of Mariano Fortuny.

WEAVING A WEB: THE LOOM

SEE ALSO

Faux Fabric
• 138

Planning Patterns
• 136

Spools of yarn (below and bottom), whether natural or synthetic, underpin the textile industry. The fabrics produced from them keep the rain off our skins, the creases in our

trousers, the stains out of our rugs, and winter's chill away from our bodies. "The web of our life is of a mingled yarn," Shakespeare wrote in *All's Well That Ends Well.* So, too, is the web of what we wear, sleep on, and drape over the windows of our homes.

A HAND loom is an ancient apparatus that interlaces, at right angles, two or more lengths of yarn or thread to form cloth. The longitudinal threads are the warp, and the crosswise threads are the filling weft (or woof). In its simplest form the loom is a single frame that is relatively easy to operate. Warp fiber, which is processed by spinning and then wrapped around a bobbin, is fixed to a cylinder at the rear of the frame and kept taut. A shuttle containing the filling yarn is passed alternately over and under the warp threads, and then, after each passage, a comblike device, called a reed, is used to push the filling thread against the previous line.

Larger hand looms are floor models having 4, 8, 12, and 16 harnesses that enable their users to weave a variety of patterns, colors, and fibers into the fabric. In the textile industry, weaving is still based on the principle of fiber crossing fiber, but the machinery that makes cloth is now completely automatic and often controlled by computer.

England's John Kay started the mechanization trend in 1733 by inventing a "flying shuttle" that was driven across the loom on a track by a lever blow, and then back. Massive power looms—first driven by waterwheels and steam—soon followed. Today, the shuttle is being replaced by innovations such as jet looms, which use pressurized air or water to propel filling thread through the weave.

Multicolored cloth (above) emerges from an automated industrial loom and grows into a large roll of fabric called a bolt. Machine-operated and hand-operated looms rely on essentially the same basic weaving processes to make fabrics out of fibers.

A weaver (upper left) fashions a kimono by hand, using warp and weft in the time-honored way. Weaving involves more than just interlacing two sets of yarn: The threading of harnesses, which determine pattern, requires experience and skill.

FROM THREAD TO FINISHED CLOTH (left): Weaving on both traditional and automatic looms requires a warp, or lengthwise thread, and a weft, or crosswise thread. Weft yarns on a bobbin shuttle swiftly back and forth between lowered and raised warp threads, while a comblike device, or reed, spaces warp yarns evenly. A cylinder at one end of the frame keeps warp thread taut, while one at the other end holds finished cloth. Looms have 2 to 16 harnesses—framelike devices operators use to weave a variety of colors, patterns, and threads into fabric.

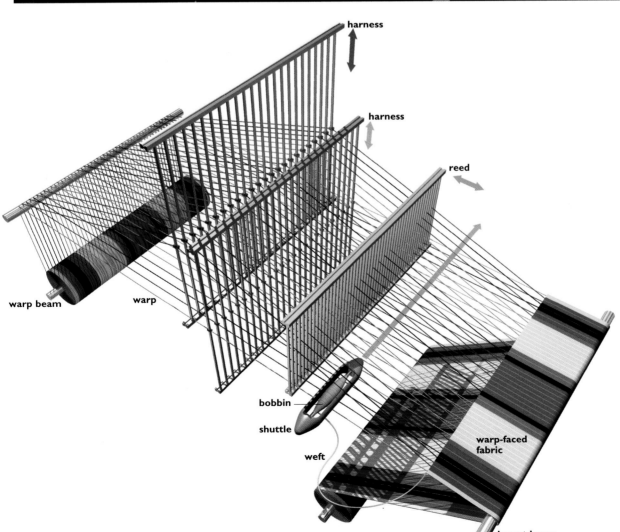

harness

harness

reed

warp beam

warp

bobbin

shuttle

weft

warp-faced fabric

breast beam

SEE ALSO

Faux Fabric
• 138

Planning Patterns
• 136

ZIPPERS

The word "zip" sounds like a fast-moving object and connotes speed. And indeed, zipping a zipper is much faster than fumbling with a row of buttons. A zipper contains two rows of interlocking teeth—metal or plastic—which are sewn on strips of tape attached to each side of an opening in a bag or an article of clothing. A zipper also has a sliding tab with wedges on each side. When a person pulls on the tab, the wedges force the rows of teeth together, closing the opening in the bag or garment. Pulling the zipper's tab in the opposite direction allows an upper wedge to force the interlocked teeth apart.

ALTHOUGH it may not be evident at first, sewing machines and knitting needles rely on the same simple principle: the loop. A machine-sewn seam holds fabric together by loops of thread, while knitting needles make rows of interconnected loops to create fabric. Sewing and knitting machines, both remarkable examples of inventiveness, do the job faster, but the idea is still the same.

The sewing machine dates from 1790, and in the mid-19th century Elias Howe, an American inventor, patented one that contained many of the features of the modern machine. Whether it is powered by an electric motor or an operator's foot on an antique treadle—the 1851 model named after inventor Isaac M. Singer had a treadle—a sewing machine needs a grooved needle threaded through an eye near the point. It also needs a thread-filled, spool-like bobbin that rotates in the machine beneath the fabric. As drive belts push and pull the needle through fabric, the thread in the needle forms a loop around the bobbin thread to make a tightened lockstitch. Another driveshaft operates a so-called feed dog that moves the fabric along.

The knitting machine was invented in 1589 by an English cleric, William Lee, to whom Queen Elizabeth refused a patent because the device was a threat to hand knitters. The machine later progressed from a simple stocking knitter to huge warp knitters, which make fabric for undergarments and outerwear out of fibers threaded from spools through guides. To select needles, the hand-operated machines use punched cards that work like the key-activation mechanism in old-fashioned player pianos. High-speed electronic machines rely on pattern software or on computers with built-in needle patterns. Knitting machines now produce an incredible range of wearables, including artificial fur.

Packed with thread, an ordinary sewing needle (left) takes on a brawny appearance under a microscope. Among humankind's oldest implements, needles date back to the bone and thorn awls made by aboriginal peoples.

Elias Howe's sewing machine (below), patented in 1846, employed the same lockstitch principle used in today's machines.

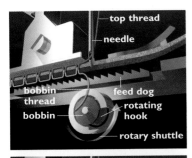

top thread
needle
bobbin thread
feed dog
bobbin
rotating hook
rotary shuttle

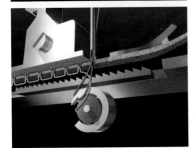

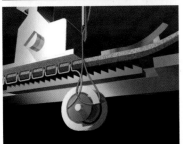

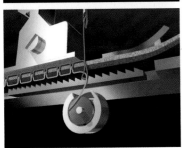

STITCHING TIME (left): The needle, thread, and bobbin work together to make simple stitches. From top to bottom, a threaded needle heads down through cloth and forms a loop of thread (red) caught by a rotating hook (blue). As the needle rises, the hook passes the loop over thread (yellow) from a bobbin (blue), and the threads join in a lockstitch. An arm at the top of the machine tightens the loop and draws the stitch up into the cloth.

Four spools of thread feed a state-of-the-art sewing machine (right). A sewing machine has a so-called feed dog (shown as green in the art at left) that advances the cloth, raises when a stitch is completed, and retracts when the needle starts down again.

SINGER

Cat. No. 2054 10-14

SINGER

Ultralock 14 U 34

SEE ALSO

Copiers • 226

Faux Fabric • 138

Lasers and Ink • 228

Weaving • 132

WEAVING a bolt of plain cloth has become a simple mechanical operation, but designing a fabric of distinction requires much creativity. More than 4,000 years ago, weavers in Peru knew how to manipulate patterns, textures, and colors to make extraordinary fabrics. Using plant and mineral dyes, they tinted yarns before weaving, alternating each strand on looms to create handsome designs, or they colored the whole fabric afterward. Both methods are in use today, although mass-production techniques such as digital color printing, intaglio roller printing, and screen printing are commonly employed.

In digital printing, a computer helps design and generate fabric prints; analog printing requires screens and printing plates. Electrostatic transfer technology may be used to create images with charged particles that attract toner—a process that is similar to one used in copy machines. Another method, called ink-jet transfer, sends streams of ink droplets to fabric mounted on a rotating drum. The thermal wax method, used for printing T-shirts, heat-fuses colored wax film onto fabric. In the method known as intaglio roller printing, designs are etched on separate rolls of copper—one for each color—and smeared with printing paste; images are transferred when fabric moves through the rolls. Designs can also be transferred to fabric by squeezing color through a stencil on a flat screen.

Distinctive patterns and effects may be woven into fabric. In plain weaves, each warp yarn or thread runs over one filling (weft) yarn, under the next one, and so on. Variations include corduroy, which has vertical ribs (wales), and basket weaves, seersucker, gauze, and twill, in which filling threads pass over one and under two or more warp threads for a diagonal appearance. Satin's sheen comes from light reflected off warp yarns passed over exposed filling yarns. Pile comes from ordinary weave that has had its filling cloth pulled out into loops that can be cut, as in velvet, or left alone, as in a towel.

Personal touch (above): A designer imprints a kimono pattern by hand, a rare sight in an era of mass-produced, computer-generated prints.

Aniline dyes (left), mainstays of the textile industry, come from coal tar, as do all aniline products, including sulfa drugs, explosives, and polyurethane.

Homespun (upper left): A bolt of classic Harris tweed takes shape on a hand loom in the Hebrides, an island group off the northwest coast of Scotland. For generations, Hebrideans have woven this durable fabric of pure wool.

Color and technology (left) combine to produce a textile pattern. Rotary printer rollers, each carrying a single color, leave their individual imprints on the fabric to produce a variegated design. More than ever, the designs of modern textiles require their creators to have a comprehensive understanding of the chemistry and technology of materials and dyes, as well as a keen appreciation of art.

FAUX FABRIC

SEE ALSO

Plastics • 190

Weaving • 132

Wonder Fabrics • 140

Wood • 180

SYNTHETIC fibers were first made in the late 19th century, when chemists learned that cellulose could be extracted from wood pulp and formed into thread through nozzles. Originally called artificial silk, the synthetic eventually was produced commercially as rayon.

Rayon is actually not a synthetic fiber; it is a reconstituted product. True synthetics are made from organic polymers, many of which soften if they are heated. Nylon, the first commercially successful synthetic, is made into clothing, rope, and parachutes. It also is cast and molded to make zippers, gears, and bearings. The range of such fibers is bewildering: Dacron, a polyester, has a nonstretch quality; polyurethane elastomers stretch; Qiana, a silklike nylon, holds a crease and remains wrinkle-free; carbon fibers reinforce; and high-strength aramid and polyethylene can resist bullets.

Most synthetic fibers are made into thread by melting the polymer and forcing it through tiny holes in a spinneret. A core construction technique spins a wrap of cotton or polyester around a continuous filament of polyester fibers, and then two or more single yarns are twisted together to form thread; another method textures the filament and heat-sets it. Air-entangled construction, used in heavy denim jeans, entangles fibers by passing them through a high-pressure air jet before twisting, dyeing, and winding. Monocord construction, used in the threads for making shoes, bonds nylon filaments together.

Dralon (above), an acrylic fiber produced in Germany, resembles soft lamb's wool when viewed at close range. Spinning turns this synthetic into yarn used to make fabrics that keep people warm.

Dralon strands (left), fuzzy from dry spinning, shimmer in the light. To make the fiber, the manufacturer forces a syruplike solution, polyacrylonitrile, through thousands of tiny openings in tall spin-pipes. Dried, collected on a spool, then warmed and cooled, the long fibers await the next step—spinning into yarn.

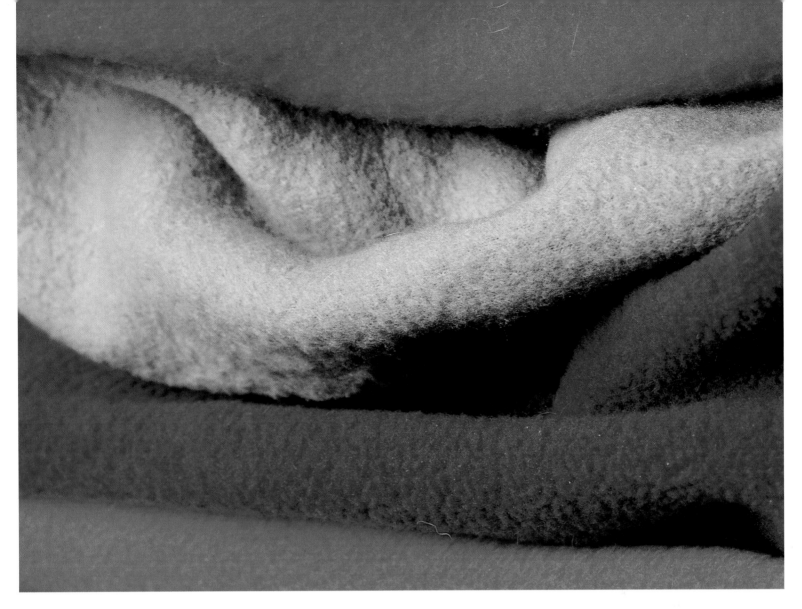

Polar fleece (above), made from the recycled polyester of disposable plastic soda bottles, keeps its wearers warm, won't shrink or stretch, and dries rather quickly. **Shetland wool** **fibers** (below left), taken from a knitted pullover, reveal scaly surfaces in an electron micrograph. A color-enhanced electron microscope image shows **nylon fibers** (below center) in a pair of low-quality, unstretched stockings. Spun by polyphemus moth larva, greatly magnified **silk fibers** (below right) look smooth and flat in polarized light.

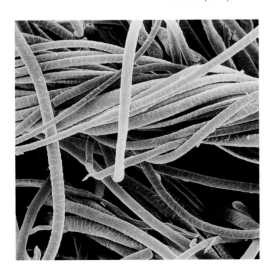

FABRICS THAT WORK WONDERS

SEE ALSO

Faux Fabric
• 138

Plastics • 190

Sensors • 42

FOR protection in battle, Persian and Roman warriors clothed themselves in overlapping metal scales attached to linen or leather. In the Middle Ages, the Crusaders wore chain mail made up of interlaced rings over their clothing. This flexible armor was effective, but it was still unwieldy, heavy metal.

Today's flak jackets and other types of bullet-resistant vests worn by soldiers and policemen were not the products of a blacksmith's shop. They were created by a new "thread science" that fabricates tough, lightweight clothing able to protect wearers from fire, cold, and water, as well as shrapnel. Fabric can be reinforced and strengthened with carbon fibers produced from simple coal tar, or by using heat to chemically change rayon or acrylic fibers. Polyester or nylon can be treated with resins that form tough, smooth protective coatings on thread. Warm, fleece-like fabric can be made from flaked and melted plastic bottles. Cloth can also be woven with glass fiber or fine metal wire.

Among the noteworthy and widely used innovations is Kevlar, a DuPont trademark for a highly crystalline polymer that is dissolved in a solvent and ultimately drawn into incredibly strong fibers used in bullet-resistant vests, military helmets, ropes, and gloves. Even more forward-looking are the "smart fabrics" used in thermal clothing. In one design—a spin-off from materials used to keep astronauts' gloved hands warm during space walks—microencapsulated "phase-change" materials, called microPCMs, do what traditional insulation methods that rely on trapping air within a fabric cannot do: They absorb and store heat, then release it in response to temperatures next to the skin.

Taking the heat (above), a chemically treated wool fabric fireproofs a garment without the health risk of asbestos. Engulfed in flame, the garbed arm's outer layer of fabric forms a heavy protective ash.

Ties that bind (opposite), the hooks and loops of the nylon fastener Velcro show in an electron micrograph. Woven separately into two pieces of fabric, they bond tightly when the surfaces of the fabric are brought together.

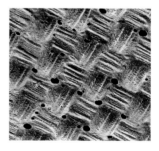

"Breathable waterproofing" weave (above) allows humid air to seep from polyurethane-coated fabric used in outdoor clothing. Surface tension and the small size of the pores keep water from entering, protecting the wearer from rain.

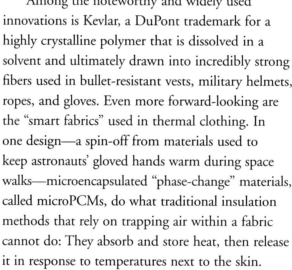

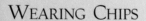

WEARING CHIPS

This cyber model sports high-tech haute couture. Outfitted as a futuristic medical courier, she wears a flexiform suit equipped with sensors, keyboard, and removable miniscreen. The future of textile wizardry is mind-boggling: Suits may change shape when mood or temperature changes; clothing may be wired with locating devices; battle-wear may warn of and deactivate toxic chemicals and germs; and camouflage fabrics may change color and pattern to match the background.

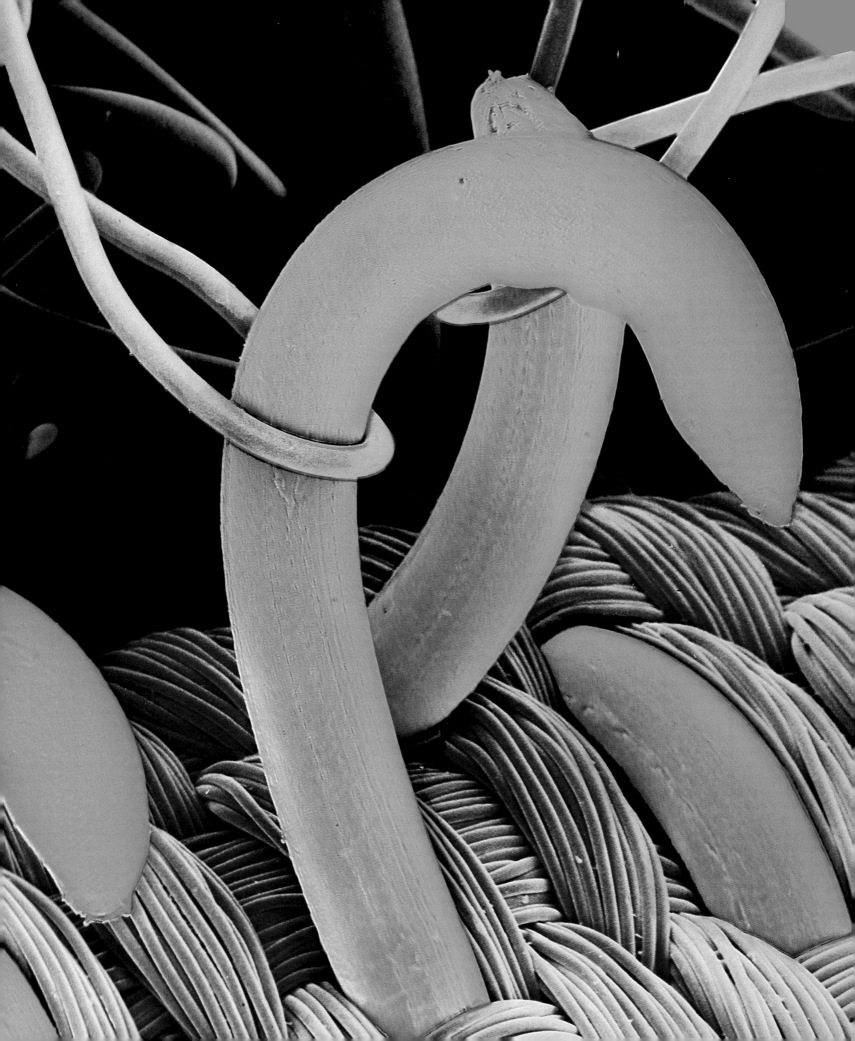

HOWEVER we define it and wherever we seek it out, entertainment serves as humankind's leavening agent and safety valve. Over the years, some of its tools have changed little in appearance and function—Chopin would have little difficulty at the keyboard of a modern piano—but the vast majority have been transformed and improved, and many more have been created. Although most of the principles have remained essentially the same, the apparatuses and the way they process information have changed a great deal. Cameras and records are now digitized. Radios come micro-size. Music can be made electronically. Video games let us sink and raise the *Titanic.* From the new, metallurgically engineered golf clubs to the ever more exciting roller coaster, entertainment's accoutrements have come under the powerful influence of technology.

Pleasing to eyes as well as ears, the flashy compact disc, or CD, symbolizes entertainment.

SEE ALSO

**Mediums and
Messages • 214**

Piano • 146

Radio • 148

MUSIC, as Thomas Carlyle said, may well be the "speech of
angels," but it is also the product of human hands, throat,
mouth, and lips. It is the result of someone manipulating the
vibrations and the increases and decreases in air pressure that
are at the root of all sound. Everything that makes sound sets
up vibrations that disturb the air and, in turn, change the air
pressure against our eardrums. The eardrums vibrate, and a
sound message is carried to the brain, which differentiates
between cacophony and an angelic choir.

Sound's loudness or softness depends
on the amplitude of sound waves, while its
pitch (highs and lows) depends on their
speed of vibration. Loud, sharp sounds, such
as rifle shots or police whistle shrieks, are
at high frequency; they create considerable
compression in the air, while a bass horn emits waves of lower
frequency. Music is composed of notes, which differ from one
another in their pitch, and what we know as a musical tone is
actually a sound of definite maintained frequency.

The instruments that play the notes create vibrations
differently. String instruments create sounds when tightened
strings are vibrated. Air blown into a woodwind vibrates a
thin, flat reed that makes the air inside the instrument also
vibrate, producing notes that can be changed by opening
and closing holes. Percussion instruments send out vibrations
when a tightly stretched skin is struck with sticks or a mallet.

Invisible to x-rays, the music
that flows from a French horn
(above) and a saxophone (left)
makes its presence known as
its vibrations stir up air inside
the instruments.

STRINGED INSTRUMENT
(below): Plucked by a player,
taut strings produce vibrations
that resonate through the
body of a guitar and become
musical sounds.

bridge

vibrating string

sound hole

Trumpet Piston Not Depressed

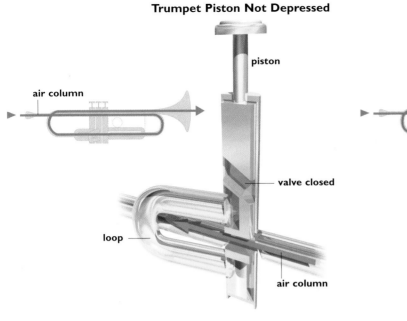

piston

air column

valve closed

loop

air column

Trumpet Piston Depressed

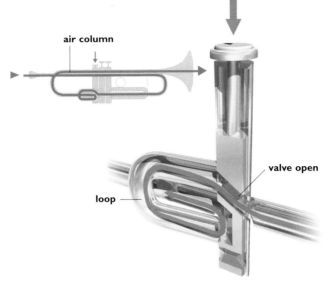

air column

valve open

loop

SOUNDING THE TRUMPET (above): Pressing and releasing a piston on a trumpet causes a valve to open and close an extra section of tubing. In the "up" position, the valve shuts off the loop attached to it, and the air goes straight through.

In the "down" position, the valve opens the loop and diverts the air through the extra section. Using various combinations of the three pistons, a musician blowing with tensed lips can create as many as eight different notes.

WIND INSTRUMENT: When a musician blows air over the mouthpiece of a flute (left and below), the air column in the instrument vibrates, producing a note. To create new notes, the flutist covers and uncovers tone holes along the length of the tube. Fingertips on fewer holes, for example, create higher notes. Unlike other woodwinds, the flute has no flexible wooden reed attached to its mouthpiece. Like some others in the group, though, it no longer comes only in wood.

no holes covered, so note is high

mouthpiece

vibrating air column

several holes covered, so note is lower

mouthpiece

vibrating air column

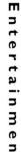

fundamental frequency

harmonics

fundamental frequency heard with harmonics

SOUND'S EFFECTS (left): Playing a musical instrument transmits vibrations that move through air as audible waves. The number of vibrations per second, or the frequency, determines pitch—high or low. Humans cannot hear sounds having less than 30 vibrations per second or more than 40,000. The simple frequency of a plucked string appears first in the group of waves shown; the next three waves depict harmonics, higher frequency multiples of fundamental waves; the bottom shows harmonics and a fundamental wave.

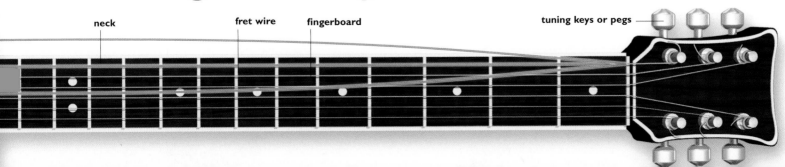

neck

fret wire

fingerboard

tuning keys or pegs

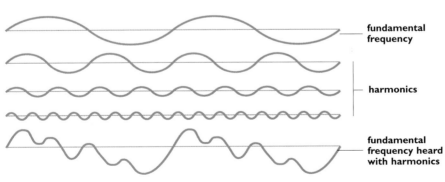

THE PIANO

SEE ALSO

Making Music
• 144

QWERTY
• 222

Radio • 148

MUSIC'S most versatile instrument may be the piano. Not only can it produce a wide range of notes, but it can also vary the loudness and duration of each one, permitting a variety of interpretations and moods. These expressive effects are possible because the strings, or wires, are struck by felt-covered hammers connected to keys that respond to the touch of a pianist.

When the earlier harpsichord was played, its strings were plucked by small picks, giving the player little control over the quality of the tone. The piano, on the other hand, could be played softly or loudly, a feature that was recognized in an 18th-century name for an early form of the instrument: pianoforte, which is Italian for "soft-loud."

The piano's action mechanism does not attach a hammer rigidly to its corresponding key. Instead, it drives it through a series of levers that achieve different tonal qualities. When the key is pressed, force is transmitted through intermediate levers called the wippen and the jack, and against a roller mounted on the underside of the hammer. This forces the hammer up to strike the string and then lets the hammer fall back immediately so that the string can vibrate. At the same time, a damper lifts from the wire. When the pianist releases the key, the damper drops back and cuts off the note.

Levers and keys (above) work pianos and typewriters. In the piano, the keys' felt hammers strike tightened strings, not ribbon and paper.

BEHIND THE KEYBOARD (below): A piano relies on lever action that begins when a musician presses the keys. The piano's operating mechanism—a complex system of levers known as the action—and one of its key components, a metal wippen, control the sounding of the strings. The levers also raise and lower the dampers, devices that stop the strings from vibrating and control the duration of the tone.

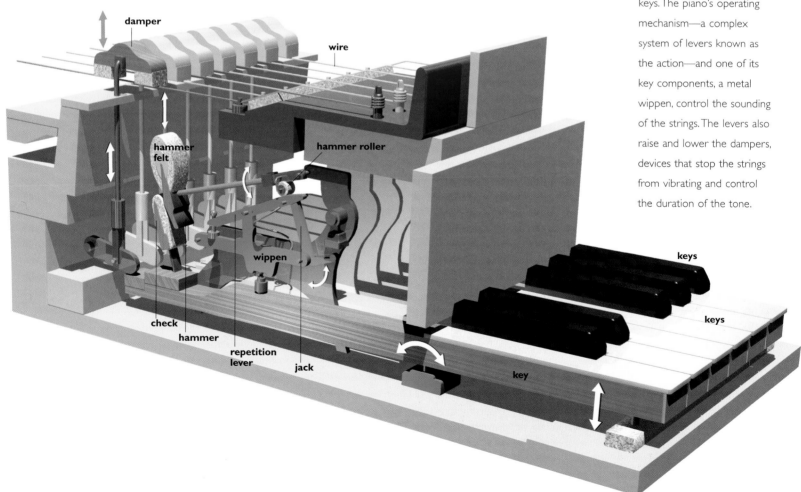

Derived from dulcimers, and preceded by clavichords and harpsichords, a splendid concert grand piano uses the same hammer action invented by Cristofori (1655-1731), an Italian harpsichord maker. With its metal frame cast and braced to resist the enormous tension of the strings, the piano has no equal among sounding boards. It produces gradations of loud and soft tones beyond the capability of the harpsichord.

SEE ALSO

Making Music
• 144

Mediums and Messages • 214

Sending Signals
• 150

Television • 156

RADIO WAVES (below) demonstrate measurable characteristics associated with their frequencies. For example, longer wavelengths have lower frequencies. Impressing sound waves onto radio waves involves amplitude modulation (AM) or frequency modulation (FM). In AM, the signal is carried by varying the amplitude of the radio wave; in FM, the signal is carried by varying the frequency along with the sound signal.

GUGLIELMO Marconi may have invented the wireless radio, but his feat drew on the research of Heinrich Rudolph Hertz, the German physicist who first demonstrated the existence of radio waves. Hertz proved that the waves could be reflected and refracted, as light is, and that they could be sent through space. He gave us hertz and megahertz, units that are used to measure the frequency of electromagnetic radiation, which includes radio waves. The units (as in *MHz*) correspond with a station's number on the radio dial and also measure the speed of computers, which require electricity.

Radio waves are made to carry signals by changing, or "modulating," the waves in some way. Some radio stations send their signals by changing the size, or amplitude, of the radio waves. These AM (amplitude modulation) stations broadcast on frequencies that are measured in thousands of cycles per second, or kilohertz. Other stations broadcast by making small changes in the frequencies of their radio signals. These FM (frequency modulation) stations are assigned frequencies in millions of cycles per second, or the megahertz range. Television makes use of both kinds of waves, with pictures carried by an AM signal and the sound by an FM signal.

In AM broadcasts, sound vibrations in the form of amplified electrical signals are impressed onto electrically generated carrier radio waves by adjusting the amplitude of the carrier waves to keep them in tune with the audio signals. Frequency modulation gives clearer transmission and reception, but it does not affect the amplitude of the carrier wave. Instead, it varies the wave's frequency in accordance with the sound to be transmitted.

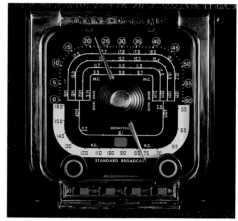

Ears on the world (above): Dial aglow, a radio lets its listeners select what they will hear by twisting a knob to a radio wave frequency.

Past and present (opposite, lower): No matter their style or age, all radios use a tuning circuit to select a desired "station," or frequency, from the many numbered choices on the band.

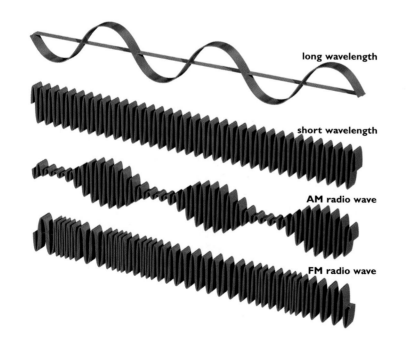

long wavelength

short wavelength

AM radio wave

FM radio wave

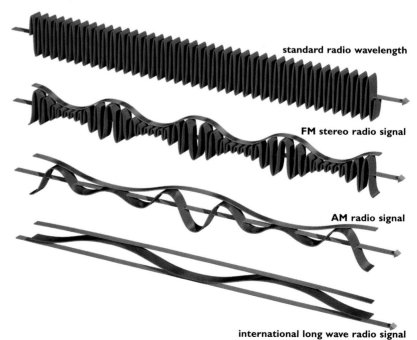

standard radio wavelength

FM stereo radio signal

AM radio signal

international long wave radio signal

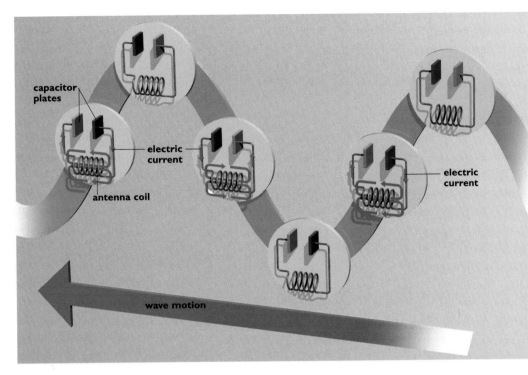

capacitor plates

electric current

antenna coil

electric current

wave motion

TUNING IN

A radio's tuning circuit picks a "station" and tunes out all the others by permitting current to oscillate at a single frequency (left). The two conducting plates of a capacitor, or condenser, store energy as electricity, while a coil to which they are linked stores energy as a magnetic field. The magnetic field collapses and sends an electric current to recharge the capacitor, which discharges again through the coil, instigating an oscillating current of one frequency. Essentially, the capacitor blocks the flow of direct current while allowing alternating and pulsating currents to pass.

SENDING SIGNALS

SEE ALSO

Mediums and Messages • 214

Radio • 148

Telephone • 212

Television • 156

THE music and the voices we hear coming from our radios have traveled great distances and in many guises. They begin, as they do in a telephone transmission, at a microphone; only this time the microphone is in a broadcasting station, where vibrating sound waves are converted into relatively weak electrical pulses. Sound waves vibrate the diaphragm of a microphone, which turns the acoustical energy into a weak electrical signal. The weak signal is amplified and then added to a carrier wave so it can be broadcast. Each radio station is assigned a carrier wave with a different frequency. An antenna at the top beams the audio-carrying radio waves—faster and more powerful than audio signals alone—at the station's assigned frequency. The distance the radio waves travel is determined by their frequency and by electrical atmospheric conditions.

After the waves leave the transmitter, they are picked up by the antenna in your radio. A tuner then selects the program by matching the receiver to the station's transmitting frequency. Weakened by the distance traveled, the radio wave signals are turned into electrical signals that are amplified and then turned into audio signals. These are sent to the radio's speaker, where the electrical waves are converted back into sound and amplified again.

Broadcasting (below): Superimposed on a quiet suburban scene, concentric circles radiate from a tall tower. Such radio transmitters typically beam powerful audio-carrying radio waves that travel in all directions and weaken with distance.

Tower of power (opposite, upper): A tower rising from the grounds of a radio and television station in Bismarck, North Dakota, bulges with electronic components and antenna configurations.

RADIO WAVES (below): When waves leave a transmitter, the path and distance they travel depend on their frequency and on electrical atmospheric conditions. At the top: Essential to worldwide communication, indirect—or short—waves reflect between the sky and Earth's surface. In the middle: FM radio and TV signals, called surface or direct waves, travel almost parallel to the Earth. They cannot pass the horizon, so relay towers extend their range. At left in the bottom illustration: Very high frequency (VHF) waves in FM, police, and citizens-band radios reflect off the ground and stay in the line of sight. At bottom right: Medium waves bounce off the ionosphere, and distant points can pick them up.

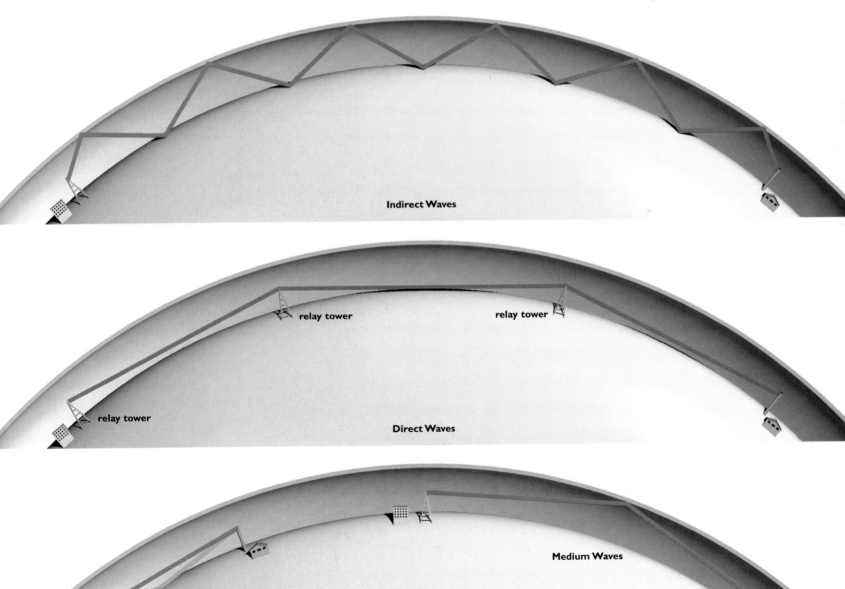

Indirect Waves

relay tower   relay tower

relay tower

Direct Waves

VHF Waves

Medium Waves

MUSICAL NUMBERS

SEE ALSO

Electronic Music
• 154

Hand Calculator
• 234

Laser • 206

Videotape
• 158

AN echo in the Alps can sound similar to human speech, but it is merely a product of sound wave reflection. Some birds that have the right vocal cords and throat formation can imitate spoken words with a fair degree of accuracy. But phonographs, tape recorders, and compact disc players, which are also nonhuman masters of mimicry, have the ability to produce a nearly limitless array of sounds with great authenticity.

In the now almost extinct phonograph, which became electrified in the 1940s, a diamond stylus, or needle, played back sound vibrations that had been scribed through a microphone into grooves on a vinyl record. As the record turned, the needle picked up mechanical vibrations, converted them into electrical signals corresponding to peaks and troughs in the sound waves, and sent them to loudspeakers that converted the signals to exact replicas of the original sound. Nowadays, sound's electrical counterparts are embedded magnetically on a specially coded plastic tape; to reconstitute the recorded sound, the tape is run past a recorder's electromagnetic playback head, which sorts out the coded sound and allows it to be amplified and sent to the speakers.

The compact disc has revolutionized music storage. Along with its computer counterpart, the CD-ROM, it requires no stylus or head to give up its sound. Instead, a thin laser beam reads the disc from underneath, probing sequences of microscopic pits and flat regions for binary codes that will be reconverted into sound of concert-hall quality.

A 4.75-INCH COMPACT DISC SYSTEM (right), spinning on its perch, outshines its predecessors—the 12-inch, black vinyl LP and belt-driven player. Inside the CD player, laser light bounces off a mirror and flashes through a focusing lens to the disc's underside, where it reads digitized sound. Reflected back to sensors, the light converts to electrical signals and then into sound.

Record-store customers in 1955 (left) regularly staked out soundproof booths to sample the latest tunes, spinning hi-fi LPs, 45s, and even the breakable 78-rpm shellac records that generally disappeared from stores in the late forties.

Wearing a stereo headset, a toddler (above) listens as tiny loudspeakers convert electrical signals into the sound of music.

compact disc

focusing
lens

lens   lens

semisilvered
mirror

beam hitting
flat surface   beam entering a pit

cylindrical
lens

beam

light sensors

AN INFRARED LASER BEAM (above) scans the track of a CD's pits and flats with pinpoint accuracy. When the beam hits a flat surface, it reflects as an "on" signal, or a 1; when it strikes a pit, the light disperses, for an "off," or 0. The ons and offs then reconvert into the sound originally digitized into the CD from a magnetic tape.

MICROSCOPIC PITS and flat areas (above), encoded with 1s and 0s representing sound, dot the reflective aluminum underside of a compact disc.

ELECTRONIC MUSIC

SEE ALSO

Computers • 236

Mediums and Messages • 214

Motion Pictures • 164

Musical Numbers • 152

Colloquially known as the amp, an amplifier (upper right) relies on vacuum tubes or transistors to increase a signal's power.

Eclectic music maker, a sound mixer (lower right) digitizes sound from numerous sources and electronically mixes and balances them before sending them to an amplifier and loudspeakers.

ELECTRONICS plays an important role in modern music making, not only in a recording studio but also in personal computers, which have built-in sound capability that outshines even fairly recent devices specialized to play music. Music, after all, is sound, and as such it can be converted by a microphone into electrical signals that, in turn, can be increased by an amplifier and reproduced by a speaker. Music can also be processed and modified electronically. It can be converted into a digital format and then blended, say, in a so-called mixer that draws on sounds from various sources and sends a balanced presentation to the amplifier. Music produced by a guitar can be turned into an electrical signal by an electromagnetic pickup device.

Music can also be produced entirely by electronics, without a conventional instrument. A synthesizer, for example, is a keyboard system of waveform generators and computer connections; it simulates the sounds and overtones of a variety of instruments by creating the appropriate electrical sound signals. The computer has expanded many human efforts and is central to today's music industry, notably in a revolutionary system known as MIDI, for Musical Instrument Digital Interface. This new system has virtually erased the boundary between the real and the artificial, allowing synthesizers to link up to other synthesizers, to computers, and to sounds from different instruments. Under MIDI's direction, an entire music studio's output can be controlled, and a single musician—a keyboardist, for instance—can become a one-person band. The future may hold "hyperinstruments" that understand the performer's intentions and enhance musical expression.

ELECTRONIC KEYBOARD

In the musical context, "keyboard" once referred only to the array of keys on a piano or an organ. No longer. While some keyboard instruments are electrical versions of the piano or organ, keyboard also means computer desktop, a workstation at which music can be synthesized, digitized, mixed, and composed (left). A vital element in MIDI, for Musical Instrument Digital Interface, the electronic keyboard—along with hardware, software, controllers, sequencers, power amps, and speakers—lets a musician design and layer sounds, dub and overdub, fill, edit, and play. In portable form, a keyboard lets a musician use it like a high-tech scratch pad that understands MIDI's arcane language.

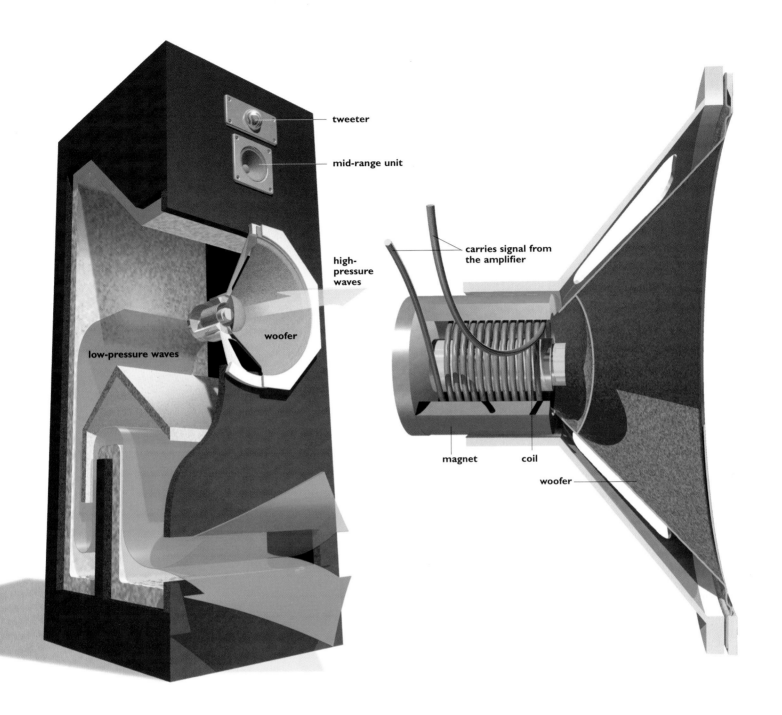

tweeter

mid-range unit

high-
pressure
waves

carries signal from
the amplifier

low-pressure waves

woofer

magnet   coil

woofer

ELECTRONIC MEGAPHONE: In the loudspeaker (above), electricity becomes sound. A typical speaker has a woofer—a large, thin, and rigid conical diaphragm that reproduces the bass part of a frequency. The mid-range unit, a smaller diaphragm, reproduces higher-frequency sounds. A coil inside a magnet in the **CONE** (above, at right) carries signals from the amplifier; the generated magnetic force vibrates coil and cone to produce sound's compression waves. High-frequency tweeters use two charged plates, one of which moves in and out under the influence of the electrical signal.

Tiny transistors (left), the power-saving replacements for large vacuum tubes, play many roles in electronic equipment, including electrical amplification and signal processing. Integrated circuits occasionally replace transistors, getting the job done with even more minuscule semiconductor wafers.

TELEVISION: THE GREAT COMMUNICATOR

SEE ALSO

Eyes and Ears
• 220

Radio • 148

Sending Signals
• 150

Videotape • 158

Laser components (above) will enhance television sets of the future. With laser units that transform electronic video signals into images of dazzling color, such sets do not even require picture tubes. Instead, they can project pictures onto virtually any surface.

Digital television (right) promises to provide much sharper images than the ones currently seen on TV, accompanied by CD-quality sound. It also will provide data and services from various telecommunications sources, including stock market quotes and e-mail. DTV's synthesis of entertainment and information may make tomorrow's TV sets much more than just magnets for couch potatoes.

AS ENTERTAINMENT and as a means of communication, television has no equal. Radio comes close, but it cannot provide the immediacy of optical images. Unlike many of the inventions and everyday technology discussed in these pages, TV was not—although the point may be argued—developed by one individual; scientists in many laboratories contributed to its development. Most of the credit, though, goes to two inventors of the 1920s. Britain's John Logie Baird developed the picture tube and was the first to televise moving objects. Vladimir Kosma Zworykin, a Russian-born U.S. engineer, developed an electronic scanning device and an electronic "television" camera called the iconoscope.

As Zworykin conceived it, a picture could be transmitted between distant points by shining it into a mica disk covered with a mosaic made of photoelectric material. The disk would be scanned by a thin electronic beam to search for weak and intense emissions. Modern television transmission borrows from Zworykin's idea and from radio transmission, which uses radio-frequency carrier waves to transport information.

Essentially, a TV camera changes light from an image into a video signal by breaking the subject into an arrangement of 525 to 625 lines that are then scanned electronically, line by line. (Newer high-definition television, HDTV, uses more than a thousand lines and a wide screen, providing movie-quality pictures.) Arriving by cable at the TV set, the sequence of pulses is turned back into scan form. It is then reconstituted as an optical image on a screen that is coated with chemicals sensitive to the three primary colors of light: red, green, and blue.

PICTURE TUBE (opposite, top): Guided by a magnetic field and fired out of three electron guns, electron beams that correspond to colors in a TV image strike millions of dots of fluorescent compound on the inside of the screen.

A TRINITRON SYSTEM (opposite, middle) uses stripes of fluorescent material and a vertical grill. It combines beams for a wide color range. Synchronization pulses in the TV signal aim the electronic guns.

ARRANGED IN TRIPLETS, dots (opposite, bottom) glow red, green, or blue as the electron beams single them out. A grill behind the screen keeps the beams from affecting other dots.

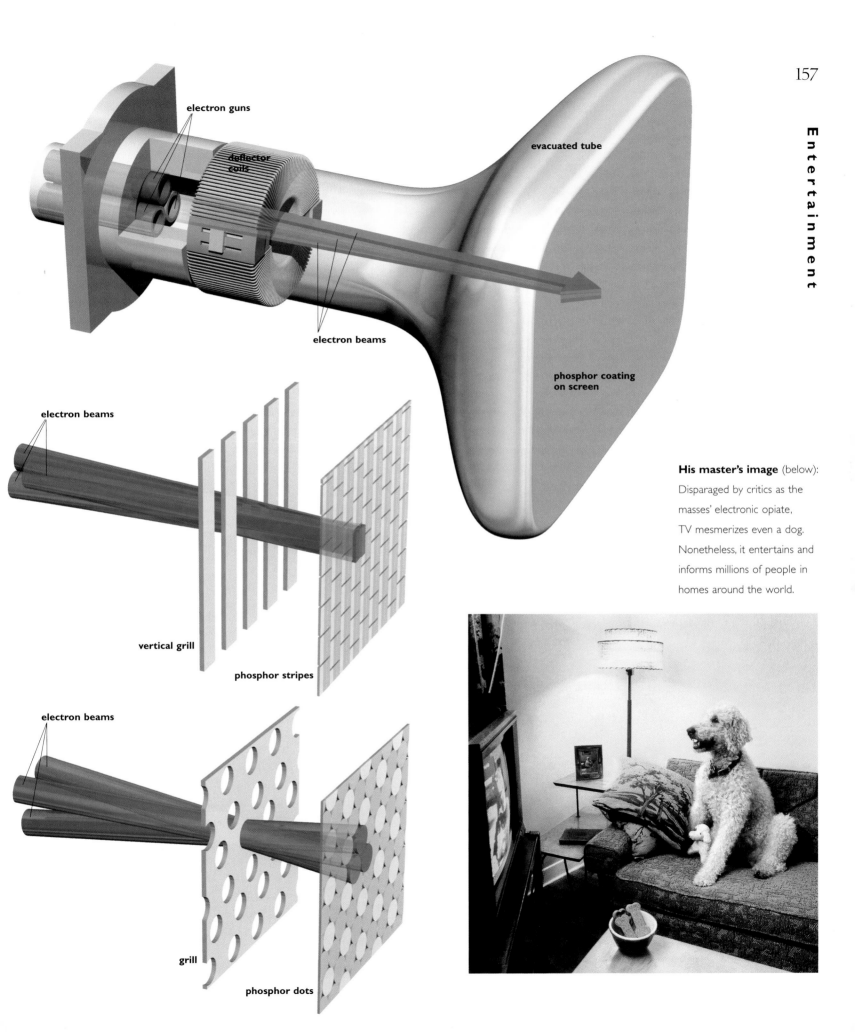

electron guns

deflector coils

evacuated tube

electron beams

phosphor coating on screen

electron beams

vertical grill

phosphor stripes

electron beams

grill

phosphor dots

His master's image (below): Disparaged by critics as the masses' electronic opiate, TV mesmerizes even a dog. Nonetheless, it entertains and informs millions of people in homes around the world.

RECORD AND PLAY: VIDEOTAPE

SEE ALSO

Camera • 160

Musical Numbers • 152

Television • 156

Video Games • 166

HOW a videocassette recorder (VCR) works is actually fairly easy to explain—the cryptic operation manuals notwithstanding. Like a television set, a VCR picks up video signals directly from a cable or antenna, but instead of projecting them onto a screen, it stores them on a reel of plastic. The reel is magnetic tape that, like its much narrower audiotape counterpart, does not need to be processed as conventional camera film.

Moving images can also be captured by a camcorder, a downsized video camera combined with a tape recorder. It focuses images onto a light-sensitive chip called a charge-coupled device (CCD) that, in turn, converts the colors, shadows, brightness, and sounds of the scene into electrical signals. The vast amount of encoded information, which requires videotape to be wider than audiotape, is passed on to tape heads and stored as patterns on a tape that is coated with magnetized iron oxide.

When a videocassette is inserted into a VCR connected to a TV monitor, the tape plays back the images, regenerating the stored electrical signals for synchronized pictures and sound on the screen. When the "record" button is pressed, TV images that flow in via the cable or antenna connection are fed into the VCR, which passes them on to tape. Looped between two reels in the cassette (as in an audiotape recorder), the tape is drawn between two sets of electromagnetic heads: One set records sound; the other set records pictures.

Turnabout (above): Behind the lens for a change, instead of in front of it, an Amazonian captures an unfolding scene with a video camera.

Tale of the tape (below): Recorder of people and events, video cameras have become as ubiquitous as still cameras. Here a videographer tapes a service in a church.

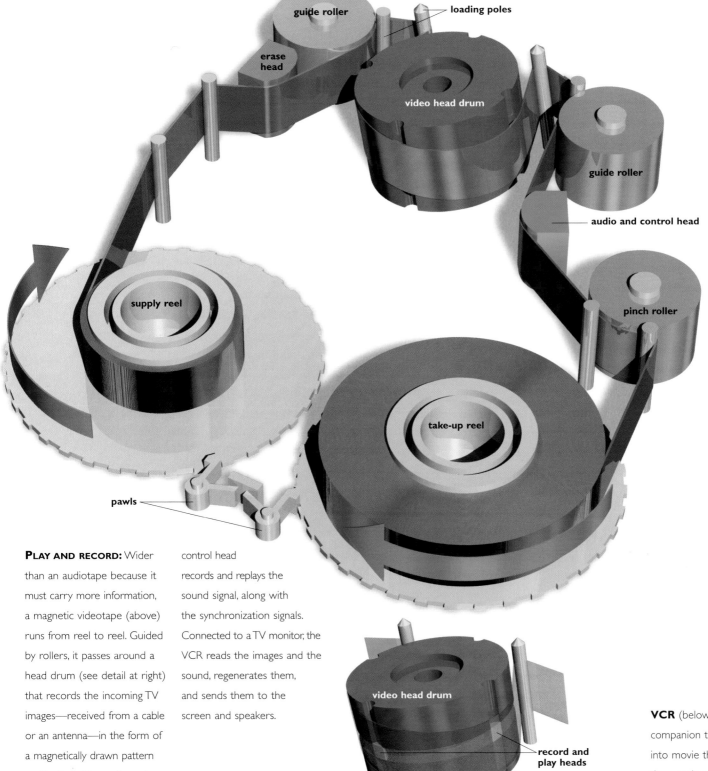

guide roller

loading poles

erase head

video head drum

guide roller

audio and control head

pinch roller

supply reel

take-up reel

pawls

video head drum

record and play heads

PLAY AND RECORD: Wider than an audiotape because it must carry more information, a magnetic videotape (above) runs from reel to reel. Guided by rollers, it passes around a head drum (see detail at right) that records the incoming TV images—received from a cable or an antenna—in the form of a magnetically drawn pattern on the tape. The audio and control head records and replays the sound signal, along with the synchronization signals. Connected to a TV monitor, the VCR reads the images and the sound, regenerates them, and sends them to the screen and speakers.

VCR (below): Television's companion transforms homes into movie theaters—minus the warnings about talking or rattling bags of popcorn.

IMAGING LIFE: THE CAMERA

SEE ALSO

Copiers • 226

From the Ground • 254

Polaroid • 162

Seeing the Light • 248

GETTING THE PICTURE:

Light gives life to the art and processes of photography. In a camera (right), a mirror and a prism correct an inverted, reflected-light scene, and a viewfinder lets a photographer see it through the camera lens, which actually consists of a system of optical lenses. Adjusting the distance between the lens and the film brings an object into clear focus. When the photographer presses the shutter button, a spring-activated device opens and closes; it keeps light out except during exposure. With the aid of a diaphragm—a fixed or adjustable component forming a large or small opening—the shutter allows the correct amount of light to come in through the lens. As the light strikes the light-sensitive film at the rear of the camera, it leaves an imprint of the scene.

ARISTOTLE and da Vinci were well acquainted with the concept behind the camera obscura: Light rays from an external object enter a darkened chamber through a tiny hole in one wall, converge and cross, and project an inverted image of the outside scene on the opposite wall. Make the room smaller and add a lens, mirrors, prism, shutter, and film, and you have a no-frills camera.

A camera is basically a lightproof container that focuses, through a lens, light from a scene onto light-sensitive film inside. Photographs are produced by light and the control of light. Too much of it means overexposed, washed-out images; too little results in underexposed, dark ones. To regulate the amount of light that strikes the film, the shutter speed is either manually or automatically adjusted; in modern cameras this normally ranges from timed exposures of several seconds to 1/1000 of a second. Cameras also have adjustable apertures that affect a picture's sharpness and contrast. Peering through the viewfinder of a standard camera, a photographer sees what was an upside-down camera obscura image that is now righted by a mirror reflecting through a prism. When the shutter release button is pressed, the mirror slides out of the way, the shutter opens, and the right amount of light bathes the film. Light exposure can also be controlled by adjusting the period of illumination. An electronic strobe, for example, flashes from a relatively slow speed of 1/100 of a second to less than 1/500,000 of a second, fast enough to catch a bullet in flight.

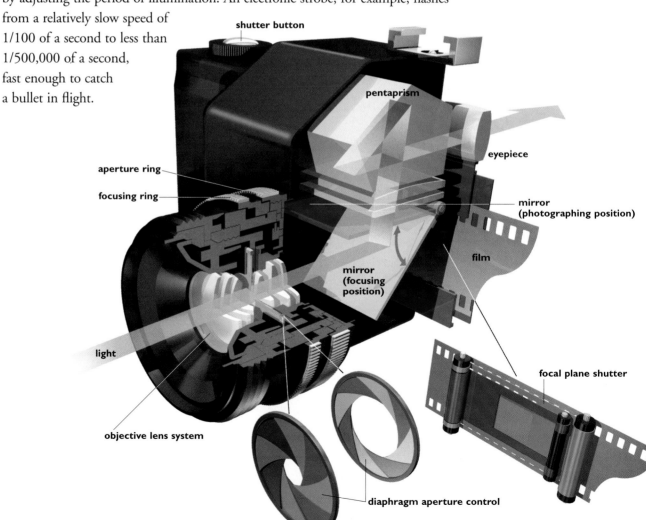

shutter button

pentaprism

eyepiece

aperture ring

focusing ring

mirror (photographing position)

film

mirror (focusing position)

light

focal plane shutter

objective lens system

diaphragm aperture control

Fixed or fitted, a camera lens (left) captures and focuses light. Elaborate cameras may have interchangeable lenses that give photographers wide-angle, telephoto, and normal views. The terms refer to the size of the visual field, or how much the final picture will show.

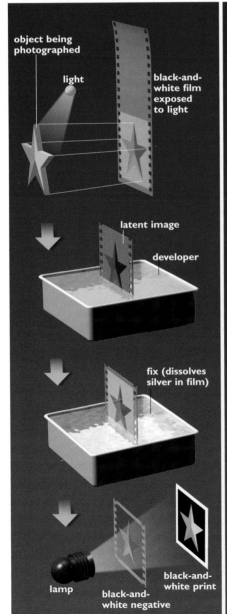

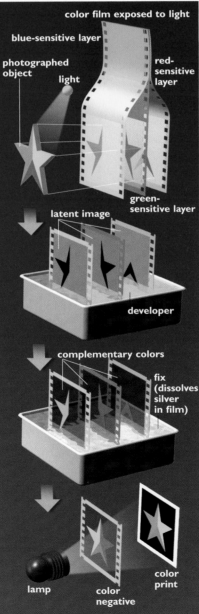

object being photographed

light

black-and-white film exposed to light

latent image

developer

fix (dissolves silver in film)

lamp

black-and-white negative

black-and-white print

color film exposed to light

blue-sensitive layer

photographed object

light

red-sensitive layer

green-sensitive layer

latent image

developer

complementary colors

fix (dissolves silver in film)

lamp

color negative

color print

PROCESSING

Photographic film is basically a strip of plastic coated with an emulsion of gelatin and crystals of light-sensitive salts called silver halides. When black-and-white film (far left) is exposed to light, the silver compound changes, and a latent image forms where the light strikes. Later immersed in a reducing agent, or developer solution, the silver turns black only in the places struck by light. A fixing solution removes the unchanged silver, making a permanent negative image. To produce a photograph, light is shone through the negative onto a light-sensitive paper. Color film (left) contains sensitizing dyes in three layers of emulsion, making each layer sensitive to a specific light color when exposed. Latent images form on the layers, and combining them brings out the true colors.

THE POLAROID AND BEYOND

SEE ALSO

Camera • 160

Copiers • 226

Electronic Music • 154

Video Games • 166

Videotape • 158

EVEN in the time of Louis Daguerre, the French theatrical designer who in the 1830s devised the first practical way of producing a permanent photographic image, camera inventors were seeking ways to produce a positive film image without having to handle a negative. In 1947, American physicist Edwin Land succeeded, giving the world the Polaroid camera, an optical system that created a finished black-and-white photograph in one minute.

Early Polaroid models work like other cameras, but their film packs contain a special positive white paper that is not light sensitive. The film packs also hold jelly-like developer-fixer reagents (stored in pouches or pods) and negative film. After exposure, the film is pressed against the paper by steel rollers; the reagents squeeze out of the pods and are spread between the film and the paper; a positive image is formed by diffusion on the receiving sheet. To produce a color photograph, the film is coated with color-responsive layers containing silver halide crystals—crystalline salts in film's emulsion that are activated by light—and layers of dyes and developers.

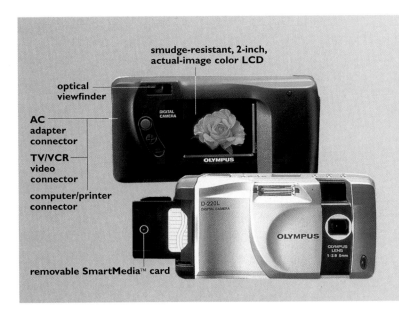

smudge-resistant, 2-inch,
actual-image color LCD

optical
viewfinder

AC
adapter
connector

TV/VCR
video
connector

computer/printer
connector

removable SmartMedia™ card

DIGITAL CAMERAS

Filmless picture-takers that rely on computers to get their images across, digital cameras have not yet replaced conventional cameras. But they are becoming more than photographic "toys." These cameras are smart in their level of electronic intelligence, as well as in their appearance. They save snapped images on a card or diskette that a computer can read. After a person has transferred the images to a computer, he or she can attach them to e-mail, insert them into Web pages, or use them to illustrate computer-generated reports.

Sun substitute (opposite, lower): Artificial light reflected from an umbrella illuminates a scene for a photographer.

Edwin Land (opposite, upper), physicist and guru of polarized light, peels an instant image from his remarkable Polaroid camera in a 1947 demonstration. Land's device used negative film, developer, and positive print paper.

Virtual reality (right): A holographic still life offers enough three-dimensional substance to tempt a viewer's touch. Holography captures a light interference pattern on a photographic plate and reconstructs it in waves that shift with movements of the viewer's head.

MOTION PICTURES: "THE MOVIES"

SEE ALSO

Camera • 160

Computers • 236

Television • 156

Video Games • 166

Videotape • 158

IN 1872, Leland Stanford bet a friend $25,000 that a running horse had all four feet off the ground at some time in its stride. When this railroad builder and former California governor asked photographer Eadweard Muybridge to help him prove it, Muybridge hit on a novel solution. He lined a racecourse with cameras, attached a string to each shutter, and stretched the strings across the track. As a horse galloped along, it broke the strings, releasing the shutters. The photographs won Stanford his bet by portraying in a series of still shots the stride of a galloping horse—something that artists had never done accurately.

Modern motion pictures are technology driven and awash in special effects, but they are nonetheless dependent on film, lenses, and single pictures. A conventional movie camera captures movement on a strip of film that is drawn past the lens aperture and stopped for a fraction of a second at timed intervals. When the film stops, the shutter opens quickly and exposes one frame, or picture; when the shutter closes, the film advances to the next frame. This sequence repeats at the rate of 24 exposures a second. Projected on a screen, a movie is an illusion of motion made possible by the fact that the human eye briefly holds onto each one of the single images until it is replaced by the next one.

MOVING PICTURES (left): In a motor-driven movie projector, a claw moves a perforated reel of film past a powerful lamp and a mirror-reflector behind the film. A shutter blocks out light while the film moves along frame by frame, from reel to reel, at exactly the same speed as the camera that took the pictures. When the film stops briefly, the shutter opens and the lens focuses the image on a screen.

FRAME BY FRAME (below): As film passes through the projector, the rotating shutter intermittently projects an image. It opens for a fraction of a second and then cuts off light by closing in front of the film.

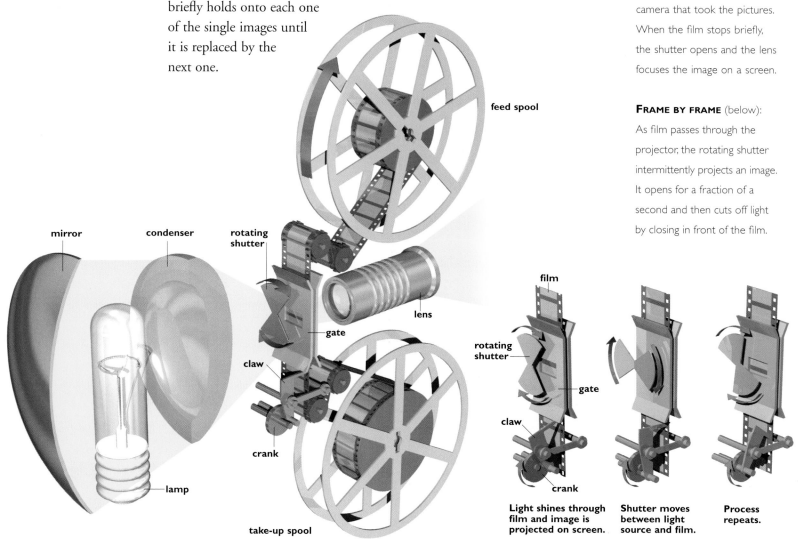

feed spool

mirror

condenser

rotating shutter

lens

gate

claw

crank

take-up spool

film

rotating shutter

gate

claw

crank

Light shines through film and image is projected on screen.

Shutter moves between light source and film.

Process repeats.

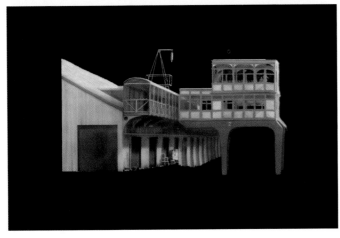

Movie magic: The *Titanic* again sets out on its ill-fated journey from Southampton, England (below), a poignant moment re-created for movie-goers by visual effects experts at Digital Domain. The illusion begins with a "plate" shot of well-wishers—extras in period costumes—gathered on James Cameron's movie set in Mexico (upper left). The Digital Domain team then filmed a 45-foot scale model (upper right) of the ship in Los Angeles and created "digital water" that they later "composited" with the plate shot. They also added digital buildings (left), birds, and smoke, and put digital people on the doomed ship's deck.

MAKE-BELIEVE: VIDEO GAMES

SEE ALSO

Backup Storage • 242

Computers • 236

Movies • 164

Musical Numbers • 152

Television • 156

WHEN a video-game player operates a mouse or a joystick to shoot down a space invader, chances are he or she is unaware of what goes on inside the computer to create the astonishing special effects on the screen. The machine may be a stand-alone console in an arcade, a compact take-home model complete with a microprocessing minicomputer inside, or a personal computer turned into a game center by binary imagery stored on a CD-ROM. Whatever its form, it is a creative collaboration of animators, 3-D graphic artists, audio programmers, interactive designers, software specialists, and electronics technicians.

Game concepts often begin with a design outline on a sheet of ordinary paper. An array of penciled connect-the-dots represents electric signals that stand for commands, screen action, background, sound, color, and movement of the images. In a process called rendering, video-game programmers "draw" images on a screen with the help of 3-D graphics microprocessors or special software programs. Animation is achieved frame by frame by using computer programs that create and play back the artwork for editing. This process is an extension of the noncomputer technique of creating, say, animated cartoons by filming a series of images hand painted or drawn on plastic "cels." Color and brightness are controlled by manipulating pixels (picture elements), the tiny bits of colored light that make up a video display. After images have been created, they are installed in a computer-generated background and stored in binary code, which can be resurrected and directed with a flick of the wrist.

Battletech (above), a 3-D video game brought to virtual life from a program stored on silicon chips, provides the packaged sound, color, and speed that characterize the imaginary world of computerized amusement.

Fantasy-driven racers (opposite), with hands on the wheels and eyes on electronic displays, line up in a Las Vegas Virtual Formula racing game.

INTERACTIVE VIDEO

Although a host of video-game critics argue that children are being tainted by the games' often violent nature, proponents respond that computerized interaction can improve hand-eye coordination, teach problem-solving, and play a role in rehabilitation therapy. In the KidsRoom (left) at MIT's Media Lab, where the walls are large video projection screens, a girl moves her arms and body to make the virtual "monster" on the screen mimic her actions. Such interactive play space gives children the ability to directly control an animation and allows them to make choices. It also assists the lab's research in computer vision and visual-action recognition.

SEE ALSO

Baseballs • 170

Faux Fabric • 138

Plastics • 190

Wonder Fabrics • 140

Headsets give coaches such as Jimmy Johnson of the Miami Dolphins (right) instant contact with players and advisors on and off the field.

Caught in midair, a fuzzy tennis ball and a lightweight racket (opposite) seem almost powerless. The ball weighs just two ounces, but its cover contains sturdy artificial fibers. Likewise, today's rackets bear only the slightest resemblance to the heavy wood, gut-strung slammers of yesteryear, but their aluminum or graphite frames actually prove stronger than wood, and their nylon strings stay grand-slam resilient.

Woods and irons no longer, golf clubs now come in space-age metals and composites. Manufacturers once cast heads in steel (right), but today they make them of strong, light, and corrosion-resistant titanium and paint them electrostatically (far right). They give them strong shafts made from graphite.

IN the past, golf clubs were made of persimmon, brass, and steel. Baseball bats were made of hickory and ash, and tennis rackets were made from maple and strung with silken gut. Football helmets were leather, and running shoes contained plain rubber and canvas. Vaulting poles were bamboo.

Today's high-tech metals, polymers, ceramics, and a host of composites put more comfort, safety, and efficiency into virtually every athletic pastime. They also increase record-breaking ability: Pole-vaulters used to manage 10 feet or so with bamboo; now they soar more than 19 feet with fiberglass poles. Some bike seats have inflatable air chambers. Lightweight tennis rackets are made of graphite, a composite material reinforced with carbon fibers. Walking shoes are shock-absorbing beauties of encapsulated EVA-foam with thermoplastic urethane heels. Even the classic birch-bark canoe has metamorphosed into a hot new item having a lightweight polyethylene hull and vinyl gunwales.

Professional baseball bats, subject to strict requirements about length and diameter, are still made of wood. College bats and other models are made of a lighter and stronger aluminum that, according to enthusiasts, will turn a .250 hitter into a .310 one. A new generation of the aluminum bat also promises to put more spring into a swing: It has a pressurized chamber loaded with nitrogen. Golf clubs, too, have improved. Wood club heads are virtually gone, replaced with aerospace titanium and sometimes steel or aluminum; shafts are made of graphite fibers bound by epoxy resins.

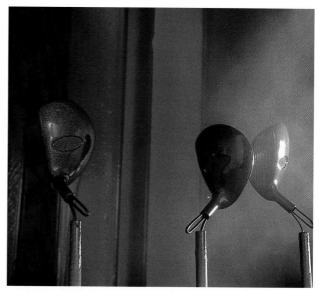

BASEBALLS AND GOLF BALLS

SEE ALSO

Mediums and Messages • 214

Old Games • 168

Plastics • 190

CONSIDER the punishment inflicted on the game ball: It may be smashed out of stadiums by Mark McGwire and other home-run hitters, bashed by a tennis racket, whacked by a polo mallet, kicked from goalpost to goalpost, and slammed around more than a hundred times during an18-hole round of golf. Still, in terms of resilience, such spheres are rivaled only by weeds, which manage to survive despite their being, as an old English lament has it, "'unted and 'acked and 'oed to death."

They manage because they are made of sturdy stuff. Baseballs, for example, are especially well insulated against blows. Each has a small core of cork or rubber in a rubber shell that is wrapped tightly within layers of wool, cotton, and polyester yarn. A latex adhesive saturates the wrapped ball, and over that is a hand-sewn, tight-fitting leather cover with 108 holes for stitches punched into it. Finally, the ball is shrunk even tighter in a dehumidifier room and sent to a ballpark where it's rubbed down before flight.

While baseballs have remained relatively unchanged over the years, golf balls have evolved from a 15th-century leather sphere stuffed with boiled goose feathers, to balls with innards made of twisted elastic bands, to today's mixed bag of aerodynamic specials with more cover choices than a magazine rack. They can be constructed of two, three, or four pieces—made up of layers of natural or synthetic rubber, resins, and sometimes thread windings—and have relatively soft, dimpled covers. The dimples on a golf ball are important because they control the distance a ball will travel as well as its flight pattern: A smooth golf ball will not travel as far as one with dimples, and its trajectory probably will not be as true.

RADAR BALL

Wired to record how fast a person pitches it, a high-tech baseball of regulation size and weight (above) has a small liquid-crystal display embedded in its side. An internal accelerometer can sense the exact moment when the baseball leaves the pitcher's hand, as well as the instant it lands in the catcher's waiting glove. Researchers can then calculate the speed of the ball based on the set distance from the pitcher's mound to the home plate.

SPORTS SPHERES (below): A baseball (at left) and a golf ball (at right) have different looks and different innards. A baseball typically contains a cork center surrounded by rubber casings, a layer of cotton yarn, and layers of wool dipped in a latex glue. A hand-sewn leather cover holds everything together. A golf ball has a liquid or solid rubber center and a dimpled cover of synthetic polymer or balata, a substance made from the dried juice of certain trees in the American tropics. The dimpling effect gives the golf ball much more than a distinctive appearance: It helps control the distance the ball will travel, as well as its trajectory through the air.

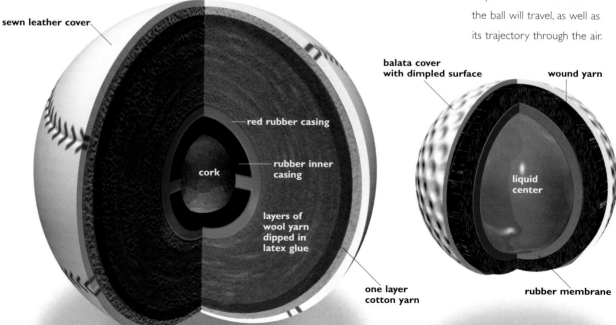

sewn leather cover

red rubber casing

rubber inner casing

cork

layers of wool yarn dipped in latex glue

one layer cotton yarn

balata cover with dimpled surface

wound yarn

liquid center

rubber membrane

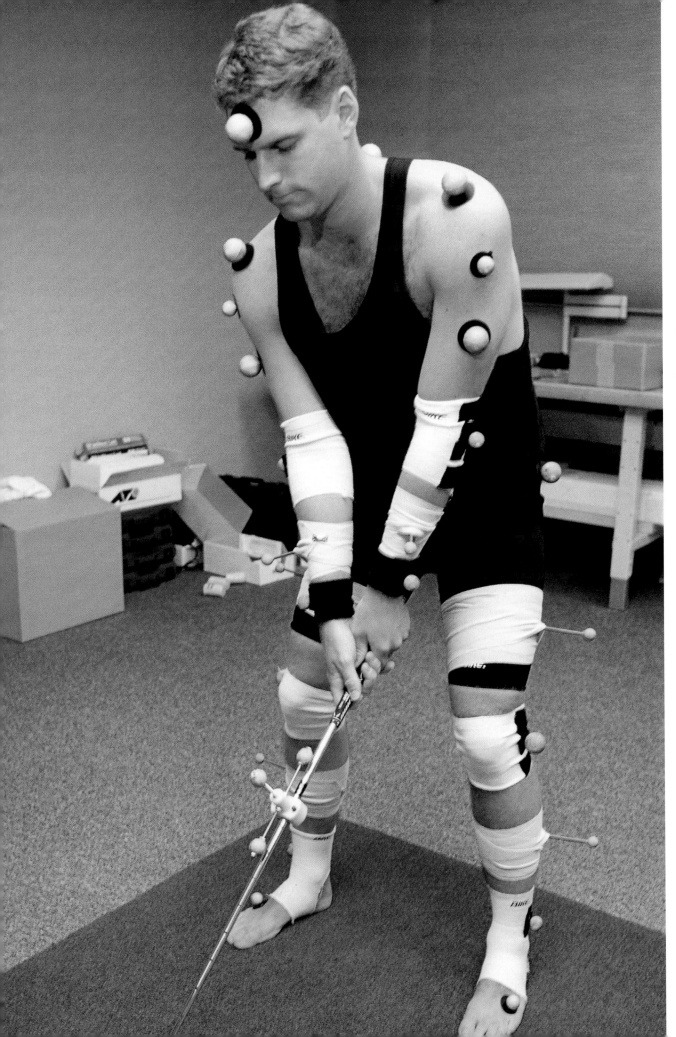

The physics of golf comes into play (left) at the U.S. Golf Association's Research and Test Center in New Jersey. When the player takes a swing, cameras record light reflected from 28 sensors on his body. Researchers will use the data to construct a three-dimensional model for measuring the force that the club places on the body's joints. The USGA requires this information before giving a golf club its seal of approval.

WHAT'S A ZAMBONI?

SEE ALSO

Fans and Air Conditioners
• 28

Grass Cutters
• 48

Washing and Drying • 34

Physicists say that ice has a very low coefficient of friction, a quality that makes it slippery enough for a person to skate or iceboat on. But when a Phoenix Coyote (right) scuffs it playing hockey, ice needs resurfacing. In the past, that meant someone had to pull a scraper behind a tractor to shave the surface. Workmen scooped away the shavings and sprayed water over the surface. They squeegeed it and then waited for the water to freeze. The Zamboni (far right) changed all that.

WHILE its name may seem to describe an Italian beverage—and it does go well on ice—a Zamboni is something quite different. It can be seen at ice rinks, and it is often as much fun to watch as the sport it serves. This innovative machine resurfaces the ice in skating rinks, and its presence at a hockey game or a figure-skating performance is certain to raise a question about how it works.

Driven around a rink, a Zamboni scrapes the rough ice surface with a 77-inch-long blade fixed at a 10-degree angle beneath the machine. The shavings are gathered up by a screw-like attachment and collected in a "snow tank" as water is fed from another tank and sent to a squeegee-like conditioner that smooths the ice. Dirty water is vacuumed, filtered, and returned to the tank, and clean hot water is spread over the ice by a flat dispenser, or "towel," which is located behind the conditioner. Zamboni drivers may not be in a class with Indy-500 racers, but on slippery ice they need a fair amount of skill. While steering with just one hand, they must operate controls that not only break up ice and snow jamming the conveyor unit but also adjust water flow, raise and lower the blade, and convey shaved ice to the dump tank.

THE ZAMBONI (right): Invented by Californian Frank Zamboni, this machine shaves time off old methods of restoring ice surfaces. While an operator drives it around a rink, its razor-sharp blade scrapes the ice surface. A horizontal screw gathers up shavings, and a vertical one propels them into the snow tank. The water tank feeds a squeegee-like conditioner that smooths the ice, while suction removes dirty water. A large "towel" spreads clean, hot water, renewing the ice.

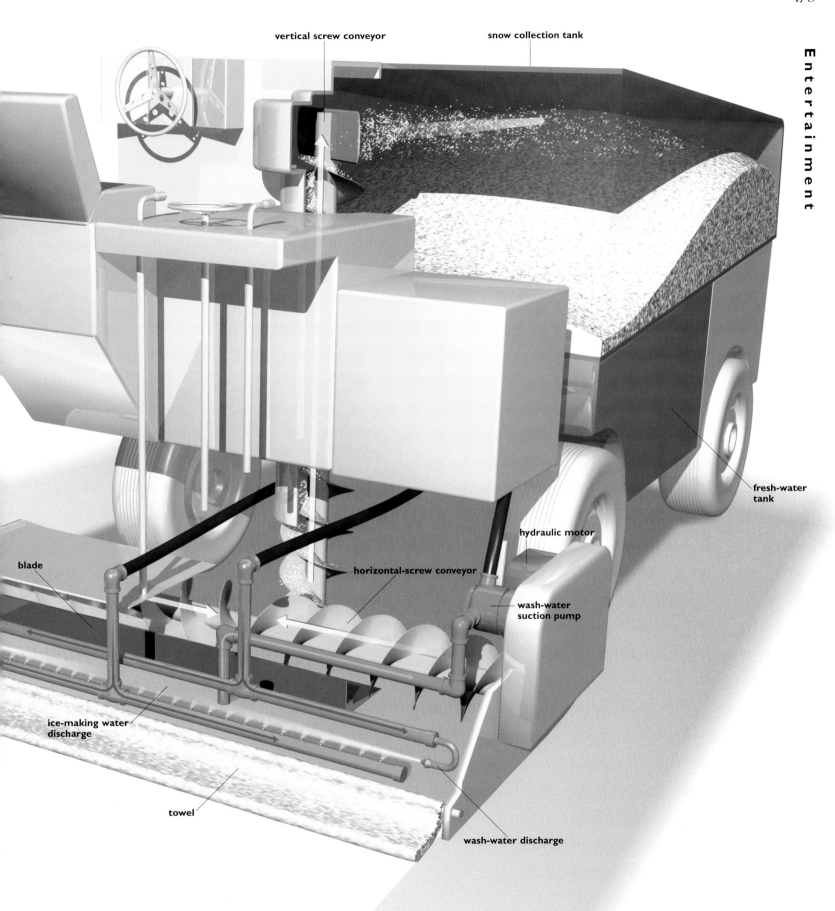

vertical screw conveyor

snow collection tank

fresh-water tank

hydraulic motor

wash-water suction pump

horizontal-screw conveyor

blade

ice-making water discharge

towel

wash-water discharge

SCREAM MACHINES

SEE ALSO

Fast Track • 96

Riding Rails • 90

Defying gravity, a roller coaster (right) hurtles through its paces in Gurnee, Illinois. Created to satisfy the human craving for thrill rides, roller coasters operate with relative safety: According to the International Association of Amusement Parks, a rider has only a 1-in-4-million chance of receiving a serious injury and a 1-in-300-million chance of a fatal injury.

THEY have intimidating names: Mean Streak, Dragon Kahn, Cyclone, and Beast. They can move you hundreds of feet up a ramp and propel you to the bottom at a 65-degree angle in a stomach-churning, bone-shaking plunge that takes only seconds. Roller coasters are the ultimate thrill ride, a mode of transportation in open cars reaching one hundred miles an hour, and they are designed for one purpose only—to give riders an adrenaline rush.

The car-trains that slowly drag you up the tracks of a wooden or steel scaffold in anticipation and apprehension, and then cause you to free-fall around pretzel-like twists and turns in pure fright, have their origin in sledlike, hollowed-out blocks of ice that raced down ice slides in 15th-century Russia. Then, as now, they fulfilled an age-old lust for heart-stopping sensation.

The first roller coaster built in the United States was conceived in 1884 by LaMarcus Thompson, who called it the Gravity Pleasure Switchback Railway. Operated at Coney Island, New York, its cars were hauled manually to the top of a 45-foot incline before being released on a six-mile-an-hour ride; passengers climbed a flight of stairs to get to the cars.

Today's roller coasters are basically still gravity machines that follow Isaac Newton's first law of motion, which dictates that a body in motion keeps moving until someone, or a force such as friction, applies the brakes. Drawn to the top of an incline by a motor and an attached chain, a roller coaster rolls along on the sheer momentum—the kinetic energy—from the first descent. The ride ends when friction and wind resistance slow the cars. In an effort to create the illusion of higher speed as the roller coaster slows down, ride engineers have added sharper curves at the end to take advantage of residual energy. Linear induction motors are being used to achieve faster speeds and help the roller coaster maintain speed as long as possible.

AGAINST THE WALL

In this decades-old photograph, several visitors to a London amusement park seem glued to the wall of the so-called Centrifuge Ride. Like the roller coaster, this popular ride also used speed to give its riders a thrill they would not soon forget. As the chamber turned, it went faster and faster, causing the people inside to feel as if they were being forced outward against the wall.

PYROTECHNIC PLEASURES

SEE ALSO

Big Steel • 186

Camera • 160

Let There Be Light • 32

Wood • 180

Handfuls of black and white ingredients (above) and others added for special effects will fill a sky with sparks, color, and thundering explosions when packed into a fireworks shell.

FIREPOWER: Rolled paper and glue hold a fireworks shell together (right). When ignited, the leader fuse burns down to a lift charge of black powder and propels the shell skyward. Inside, a time-delay fuse causes the color-producing chemicals to ignite and scatter; it also sets off the noisemaking salutes. For full effect, pyrotechnists time shells to explode at the highest point of their trajectory.

FIREWORKS have elicited oohs and aahs from delighted watchers ever since the ancient Chinese gave the world an early form of rockets, Roman candles, and fiery pinwheels. The Chinese may have invented pyrotechnics for use as weapons or to scare away demons, but not much time went by before they realized that fireworks would play very well as outdoor entertainment.

Even in the tenth century, fireworks technology was not so different from that of today; it relied on gunpowder mixed with various chemicals to provide color, and it included metal shavings that would create the sparkle effect. Fireworks now are generally tube-shaped or spherical paper containers packed with explosive powder and a time-delay fuse. The containers also hold strategically placed packets of metallic salts: Lithium or strontium produces red; barium nitrates make green; copper compounds result in blue; sodium creates yellow; charcoal and steel produce sparkling gold; and titanium makes white. Powdered iron, aluminum, or carbon produce the sparks and other special effects.

A pyrotechnist places a fireworks shell in a mortar or a special gun and lights the shell's main fuse, using either a road flare or a computer-controlled electric switch. The lift charge ignites and propels the shell high into the sky. Next, the time fuse ignites the various compounds, creating patterns and shapes based on the placement of the chemicals within the shell. Of course, the imprint in our mind's eye lasts much longer than the life of the shell and its display.

Brilliant blooms of floral fire (opposite) paint the sky as a time-delay fuse detonates fireworks in a carefully timed, multicolored light show.

Don't try this at home: A pyrotechnics packer in Japan (above) carefully loads a container with powders and metallic salts. Later, someone will use a mortar to send his handiwork high into the air.

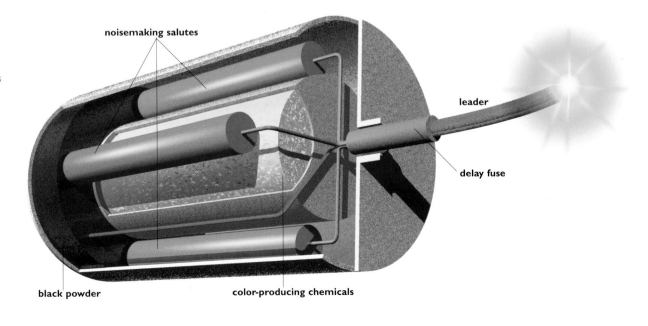

noisemaking salutes

leader

delay fuse

black powder

color-producing chemicals

MINING

"IF you have great talents, industry will improve them," English portrait painter Sir Joshua Reynolds once said. "If you have but moderate abilities, industry will supply their deficiency." His remarks came during a 1769 address to students of the Royal Academy, and his use of "industry" referred to diligence in any pursuit. But it also could have applied to the industrial revolution, which introduced power-driven machinery to late 18th-century England and forever changed the workings of the factory system. Today, the industrial world employs the modern tools of technology and science to manufacture products, extract value from natural resources, and create new versions of the tools themselves. Increasingly, manufacturers are relying on information technology to help them improve efficiency and become more competitive in the global market.

Born in industry's fiery furnace, molten steel announces itself in a glowing shower of sparks.

WOOD: ALL-PURPOSE RESOURCE

SEE ALSO

**Construction
Elements • 66**

**Cutting Edges
• 46**

**Faux Fabric
• 138**

Wood pulp (below): When smoothed and rolled, these rough and tangled microscopic fibers will become crisp sheets suitable for front-page news.

Wrinkled ends (right) of a giant roll of paper overwhelm a trimmer at Fraser Papers Inc., in Madawaska, Maine.

Felled trees (far right) become mountains of logs at a lumber mill in Oregon. Transformed by industry— pulped for paper, dressed into boards and timbers, and processed into rayon—wood and its products help clothe, house, warm, and inform us.

ACCORDING to an old saying, he who splits his own wood warms himself twice. This adage seems a major understatement if you consider the things lumber does for us. In the United States, for example, we take more than 30 billion board feet of lumber from our forests to build 95 percent of our homes. The average citizen uses about 750 pounds of wood-derived paper each year, while many thousands of tons of cellulose-based textiles and chemicals are used throughout the world. Without wood, our planet might not even be environmentally sound: For every ton of wood a forest grows, it removes 1.47 tons of carbon dioxide—which in excess contributes to global warming—and replaces it with 1.07 tons of oxygen.

Industries and individuals own more than 70 percent of the 737 million acres in U.S. forests, and the net growth in those forests exceeds harvesting by 33 percent if you combine the number of naturally regenerated trees with that of seedlings planted each day. The sawmill industry depends on harvesting, which is the starting point for the many conversions wood undergoes. After timber has been logged and debarked, it is sawed into boards that are trimmed and edged, graded and dried, and planed and treated with decay-preventing chemicals. Steam-heated logs are spun on lathes and skinned to make veneers that, in turn, are glued together to form plywood. Particleboard and fiberboard are made from wood residues mixed with resins and wax. Wood chips cooked into pulp make paper, and digested cellulose eventually becomes rayon, proving that "wooden" doesn't always mean stiff and lacking flexibility.

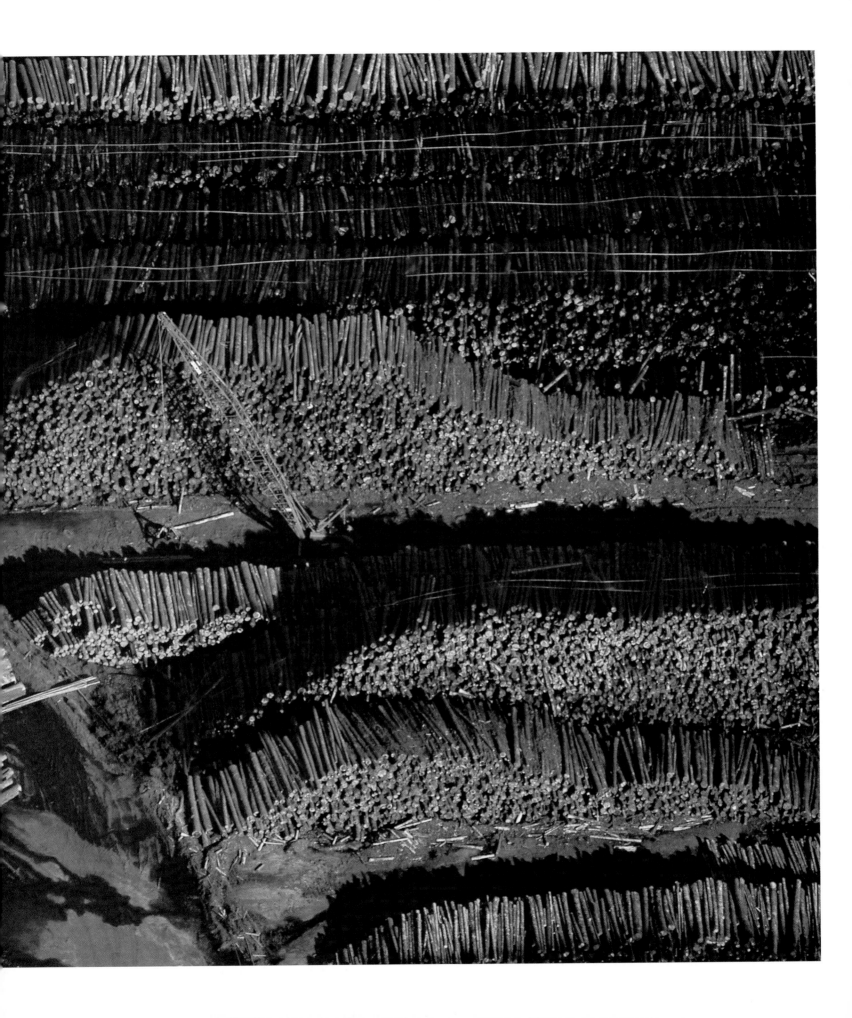

GLASS: THE RIGID LIQUID

SEE ALSO

Plastics
• 190

Recycling
• 192

Solar Heating
• 26

WE peer into glass to see ourselves and through it to see others, for this material is both reflective and transparent, depending on how it is treated. Its basic composition can be reduced to three essential ingredients melted together at high temperatures: sand, soda, and limestone. In more elaborate chemical terms, this translates into silicates and an alkali flux, with metallic oxides added for color.

Glass is similar to most other solid materials except on a microscopic level. There, it lacks the orderly molecular arrangement of true solids, a disorder that makes it resemble a liquid. Indeed, it has been termed "the rigid liquid," a reference to its high viscosity—the property of a fluid by which it resists shape change or relative motion within itself. Glass is transparent because its atomic arrangement does not interfere with the passage of light. (A glass mirror reflects because a thin layer of molten aluminum or silver was applied as backing.)

The actual manufacture of glass dates back to 3000 B.C., when Egyptians glazed ceramic vessels with it. Much later, the Romans made glass for utilitarian and decorative purposes. The art of stained glass flourished in the Middle Ages. Today, glass is made in large crucibles in furnaces where the melting temperature reaches 2900°F. Skimmed of impurities and cooled, molten glass may be poured into molds and pressed, blown, cast for mirrors and lenses, or "floated" or drawn to produce window glass or tubing. Shaped glass goes through an annealing process to fix colors and to remove internal stresses and make it less brittle. Depending on its intended use, the cooled glass is then ground, polished, bent, laminated, or decorated. Some products that require high strength, such as glass doors or eyeglasses, are specially tempered. This rapid cooling process is the reverse of annealing in that it induces high, permanent stress.

Red-hot and new, glass bottles (below left) take shape on a production line. Except for automation and new materials, the manufacture of glass bottles has changed little since people in Egypt and Mesopotamia made glass vessels many centuries ago.

At Corning Incorporated (below), fire polishes laminated plates. Windshields for cars and aircraft contain laminated glass, made by bonding layers of glass with resins and then compressing them with heat.

Molten glass (opposite) pours from a furnace, ready for cooling, pouring, molding, pressing, blowing, or drawing.

KING COAL

SEE ALSO

Liquid Gold
• 188

**Making
Electricity** • 54

Riding Rails • 90

Tunnels • 76

THE world's vast stores of coal were probably laid down some 300 million years ago during the Carboniferous period, when the remains of plants decomposed and compressed to form the hard, black substance we burn as fuel. The U.S. alone has some four trillion tons of this organic rock, and it is enough to take us, at the current rate of consumption, well into the 24th century.

Like oil and gas, coal must be extracted from the ground. One way this is done is by excavating outcrops of coal deposits through strip mining, a process that removes surface material to expose coal seams or beds. The equipment is mammoth: A typical truck's bumper might be as high off the ground as a worker's helmet, and the load carried may be about 250 tons a trip.

Underground mining is the other major method of extraction. Entered by tunnels or shafts dug into the ground, such mines once were supported solely by wooden timbers. Miners, paid by the tons of coal they loaded, had to muscle out their coal carts. Today, new roof-bolting technologies are used, and mines are drained, ventilated, and fitted with mechanical and electronic equipment.

A typical underground mine is excavated in several ways, but the most common is the room-and-pillar method. Rooms generally 20 to 30 feet wide are cut into the coal bed, leaving huge pillars, or columns, to support the roof and control the airflow. The cutting is done by a machine called a continuous miner, which obviates the need for blasting and drilling. When the cutting reaches the end of the line, workers begin "retreat mining" in which they pull as much coal as possible from the remaining pillars until the roof collapses.

On a train ride to a sunless world, coal miners (above) enter a tunnel without having to stoop, as their predecessors often had to do. In the past, miners worked long hours, and the tonnage they loaded determined their pay.

HYDRAULIC SUPPORTS (opposite, lower right) prop up the roof of a coal face, preventing its collapse.

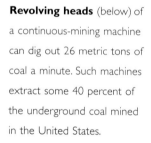

Revolving heads (below) of a continuous-mining machine can dig out 26 metric tons of coal a minute. Such machines extract some 40 percent of the underground coal mined in the United States.

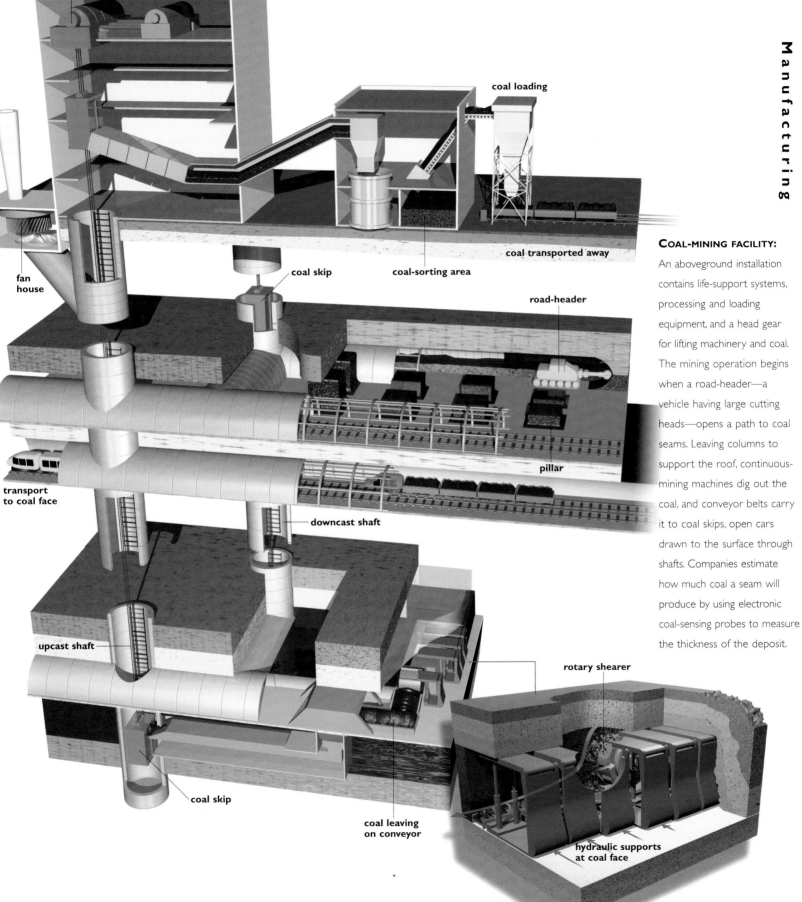

head gear

coal loading

coal transported away

coal-sorting area

coal skip

fan house

road-header

pillar

transport to coal face

downcast shaft

upcast shaft

rotary shearer

coal skip

coal leaving on conveyor

hydraulic supports at coal face

COAL-MINING FACILITY:
An aboveground installation contains life-support systems, processing and loading equipment, and a head gear for lifting machinery and coal. The mining operation begins when a road-header—a vehicle having large cutting heads—opens a path to coal seams. Leaving columns to support the roof, continuous-mining machines dig out the coal, and conveyor belts carry it to coal skips, open cars drawn to the surface through shafts. Companies estimate how much coal a seam will produce by using electronic coal-sensing probes to measure the thickness of the deposit.

BIG STEEL: FROM ORE TO STAINLESS

SEE ALSO

Boats Afloat
• 102

Construction Elements • 66

High Steel
• 70

Recycling • 192

FEW industries can match the massive, complex equipment and the awesome spectacle of steelmaking. From raw and dirty lumps of iron ore to sheets of shiny, corrosion-resistant stainless steel, the process is a meld of intense heat, blasts of high-pressure air, molten metal, violent boiling and bubbling, arcs of electricity, and the din of forges and rollers. First mass-produced in the 19th century, steel is a mainstay of ships, automobiles, and skyscraper frames.

If the process of steelmaking could be reduced to a few terms, those terms would include the blast furnace, Bessemer converter, and open-hearth furnace. Steel is an alloy of iron and a tiny amount of carbon and other elements, a mix that causes the iron atoms to bind together tightly and produce a material that is even harder and tougher than iron. To create steel or to make cast or wrought iron, the iron is first extracted by smelting oxide ores mixed with coke (a form of carbon) and limestone in a blast furnace—a towering, cylindrical stack where a blast of air fuels combustion. In molten form and full of impurities (pig iron), it can be made into cast or wrought iron. To convert the pig iron to steel, it can be refined in a Bessemer converter, which uses hot compressed air to remove impurities. An open-hearth furnace may be used instead.

Stainless steel, a long-lasting alloy containing at least 10 percent chromium, is generally produced in an electric-arc furnace. Carbon electrodes melt iron so that it can be mixed with stainless steel scrap, chromium, and other elements, such as nickel and molybdenum. The mix is cast into ingots or a slab and then hot-rolled, cold-rolled, or forged into final form.

Molten steel (below left) gets spot-checked by workers. The chemical composition of alloys differs to meet various needs, such as corrosion conditions, temperature ranges, and strength requirements. Manufacturers work steel into many shapes, including **coils** (below). They can cold-roll it to reduce the thickness of its sheets or draw it into rods and wire. Steelworkers can also cast it into slabs, billets, and blooms.

White-hot molten iron (left) seethes at a temperature of 2800°F as it pours from a cauldron into a steelmaking furnace. Manufacturers can employ Bessemer furnaces to convert iron to steel, or they may rely on electric-arc or open-hearth furnaces.

STEELMAKING (below): When workers dump iron ore, limestone, and coke into a blast furnace, the coke and blasts of air fuel combustion. From the furnace, molten pig iron goes to a converter charged with oxygen to burn out impurities and convert the iron to steel. The next step involves removing the molten steel and tipping the converter to drain off slag.

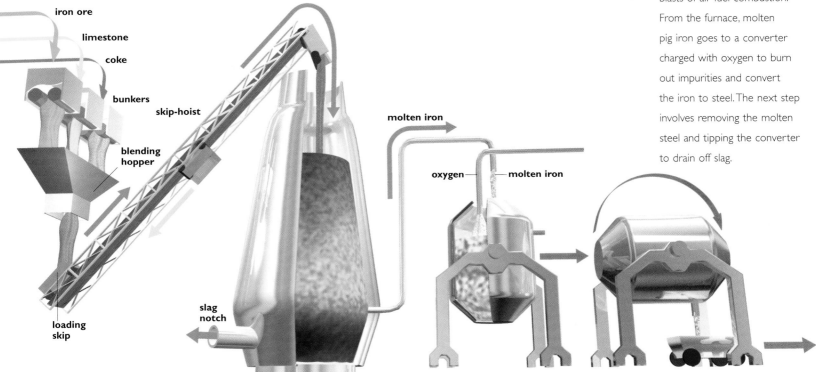

iron ore

limestone

coke

bunkers

skip-hoist

blending hopper

loading skip

slag notch

molten iron

oxygen — molten iron

blast furnace

oxygen furnace where oxygen mixes with iron to make steel

steel poured off

LIQUID GOLD: DRILLING FOR OIL

SEE ALSO

Alternatives
• 86

Home Heating
• 24

**Making
Electricity • 54**

Plastics • 190

A COMPLEX mixture of hydrocarbons that formed from the organic debris of long-dead plants and animals, petroleum is the oily, flammable liquid of power and petrochemicals. Pooled deep in the ground, within layers of rock under intense pressure, it is extracted by a drilling rig, a series of rotating pipes supported by a derrick. When a reservoir underground or offshore is tapped, oil tends to burst out explosively, a tendency that used to provide scenes of black gushers spewing into the air. But today's technology and environmental sensitivity have made such sights a memory. Drilling rigs—which probe to depths as great as 25,000 feet—pump a flushing mud into the ground to carry debris to the surface and prevent oil from erupting.

After a well is drilled, the oil usually flows up under its own pressure and into a separator, where gravity helps remove the oil from the mix of briny water, natural gas, and sand that came along with it. The cleansed crude oil then is sent through an array of pumps, compressors, and dehydration towers to make it suitable for refining; at the refinery, petroleum undergoes fractional distillation, which produces gasoline, kerosene, diesel and fuel oils, lubricants, and asphalt.

In the early days of prospecting for oil, a reservoir often was discovered by chance when, say, a farmer spotted oil mixed with the water in his well. Nowadays, ultrasound waves provide color-coded seismic data that show images of rock formations in the ground or seabed. When oil is found, it is transported far more easily than in the past: Each day, for example, the Alaska pipeline brings a million barrels of crude oil over 800 miles of wilderness to tankers that can carry several thousand metric tons apiece.

FRACTIONAL DISTILLATION (opposite) refines crude oil by using the distinctive boiling and condensation points of different hydrocarbons to separate them out. Crude oil vapor enters at the bottom of a fractionating tower. The heated vapor rises, cools, and condenses on trays at different levels—the lightest fractions with the smallest molecules at the top. Catalytic cracking in a reactor breaks large molecules into more valuable small ones. A vacuum still lowers the boiling point of the heaviest, unvaporized oil.

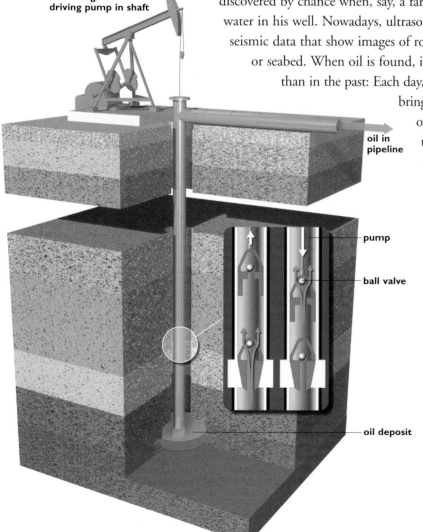

beam engine
driving pump in shaft

oil in
pipeline

pump

ball valve

oil deposit

PUMPING OIL (left): Activating a system of ball valves near the bottom of an oil shaft, a beam engine dips and rises to lift oil from porous rock 500 to 25,000 feet deep in the ground. At the surface, pipelines carry the oil to processors.

Bright lights (right) and dark towers at a petrochemical plant dominate a landscape.

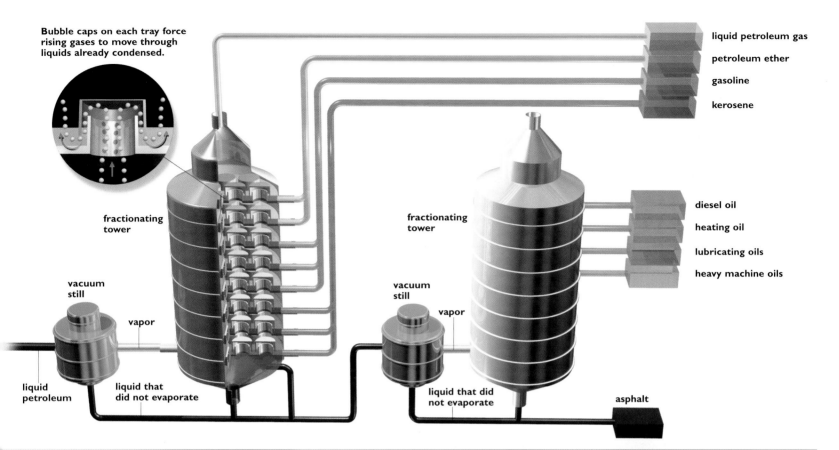

Bubble caps on each tray force rising gases to move through liquids already condensed.

liquid petroleum gas

petroleum ether

gasoline

kerosene

fractionating tower

fractionating tower

diesel oil

heating oil

lubricating oils

heavy machine oils

vacuum still

vacuum still

vapor

vapor

liquid petroleum

liquid that did not evaporate

liquid that did not evaporate

asphalt

SEE ALSO

Faux Fabric
• 138

Glass • 182

Liquid Gold
• 188

Recycling • 192

Wood • 180

IF wood is one of nature's great genuine articles, plastic is the welcome, artificial usurper, a take-charge material with a seemingly endless capacity for changing its shape and function. Most plastics are by-products of petroleum or coal, which are organic materials, but they are considered artificial because they are made rather than grown.

Plastic came into being more than a century ago when a $10,000 prize was offered to the person who could create a substitute for ivory in billiard balls. John Wesley Hyatt of New York, an inventor-entrepreneur, entered the competition, and although he did not win, he came up with semisynthetic Celluloid, a mixture of nitrocellulose and camphor put under heat and pressure. Hyatt's invention was a commercial success that found immediate use in dental plates, eyeglass frames, combs, and men's collars.

Today, more than 50 varieties of plastic stand in for a host of materials, and it is hard to imagine anything spun or molded without a plastic presence. As Lucite or Plexiglas, plastic substitutes for glass in aircraft windows, car taillights, boat windshields, clock faces, and camera lenses. As Bakelite, it replaces rubber as an electrical insulator. As Corian it passes for marble, and as Styrofoam, for insulation and disposable food containers. Polyvinyl chloride (PVC) substitutes for metal in drainpipes; Teflon keeps food from sticking to skillets; nylon and other synthetic fibers replace wool and cotton.

Plastics generally start out as powders of polymers, the binder in a mix that includes plasticizers, fillers, and pigments. The raw materials can be compressed in molds under heat and pressure or poured into cold molds to harden. They can be squeezed between rollers or extruded through a die to be cut in lengths or coiled. Because they do not decay, plastics present environmental problems. Often maligned, they are nonetheless remarkably useful creations that are here to stay, for better or for worse.

PRESSURE COOKER

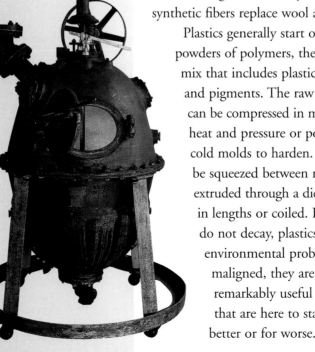

In the clunky contraption shown at right, the Belgian-American chemist Leo Hendrik Baekeland mixed up a batch of what would become Bakelite, the first synthetic polymer. Baekeland's creation was a blend of the disinfectant carbolic acid (phenol) and the preservative formaldehyde. This hard, clear solid was impervious to acids, electricity, and heat. Molded and tinted, it was made into a number of products, including radio cabinets, buttons, knife handles, telephones, and airplane parts.

Elastic and tough, a sheet of petroleum-based polyethylene (above) takes shape, ready for fashioning into garbage bags and textiles. This marvel of polymer science has a durability that makes it difficult to dispose of in town dumps.

Plastic items may surround and even dwarf us (opposite). Another invaluable by-product of the oil industry, plastic has a wide range of properties: hardness, transparency, elasticity, toughness, insulating capacity, and resistance to corrosion. The thermosetting class of plastics does not allow resoftening or reworking. The thermoplastic class, on the other hand, will soften or melt when heated.

Polyethylene fabric (below) reveals its netlike structure and tensile strength when seen under an electron microscope. Manufacturers create the polyethylene fibers from polymers, large molecules made by linking smaller ones.

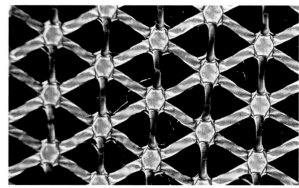

Pellets of plastic (above) resemble a more ephemeral substance: ice. Manufacturers can mold and otherwise fashion plastic into endless shapes for a variety of needs.

ALL THAT WASTE: RECYCLING

SEE ALSO

Big Steel • 186

Faux Fabric • 138

Glass • 182

Plastics • 190

ACCORDING to the National Solid Wastes Management Association, Americans generated approximately 210 million tons of solid waste in 1997. Each year we throw away billions of batteries, razors, blades, and disposable diapers, along with millions of tons of paper, glass, plastic, leaves, and grass clippings. If we were somehow able to place all our discarded beverage cans one on top of another, they would reach the moon almost a score of times.

Fortunately, recycling is helping. It can be done in a number of ways, such as reusing materials left over after a manufacturing process. Steel scrap, for instance, can be used to make stainless steel, or glass scrap may be remelted to make more glass. Consider the cutting and shaping of plastic sheets, a process that leaves much material behind. One Japanese manufacturer, Mitsubishi Rayon, handles 2,000 tons of waste plastic a year, using a heating process to break down acrylic plastic from a complex polymer to a simpler form that becomes the raw material for more acrylic plastic. The process can convert as much as 85 percent of scrap plastic into its reusable, basic form.

Another method reclaims materials from worn-out items, using cardboard boxes and newspapers, for example, to make business cards, paper towels, and more newspapers. Lead batteries can be made into new batteries; steel cans can be used to make more steel; discarded glass can be remelted to make new bottles and fiberglass; plastic makes fiberfill for ski jackets or is reshaped into traffic cones. Precious metals such as gold, silver, palladium, and platinum can be recovered from printed circuit boards, plated and inlaid metals, and photographic wastes.

Waste with a purpose: Chipped plastic soda bottles (above) get recycled into toys, traffic cones, carpet backing, fiberfill for ski jackets, and even warm, fleece-like fabric.

Crushed cans (below) at Alcoa's Recycling Center in Tennessee will soon be part of new cans. Making beverage containers from scrap requires only about 30 percent of the energy needed to make cans from primary metal.

Sorting and sifting, a recycler (opposite) gets something back from things thrown away. Recycling reduces the enormous expense of disposal and incineration; it also provides a new resource.

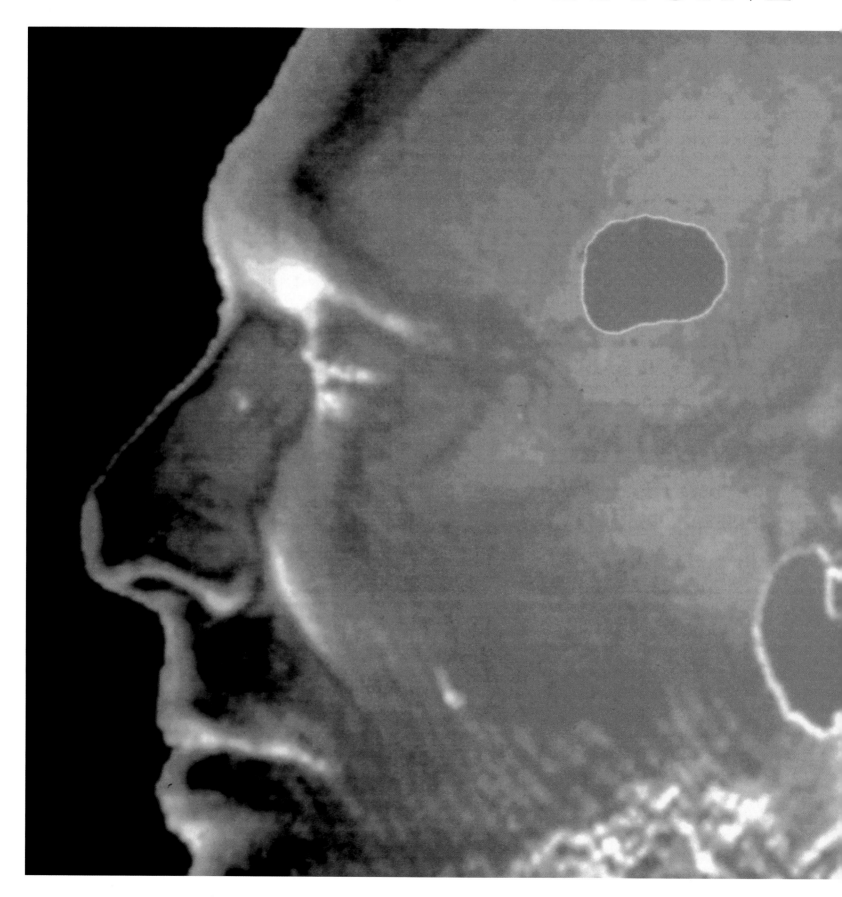

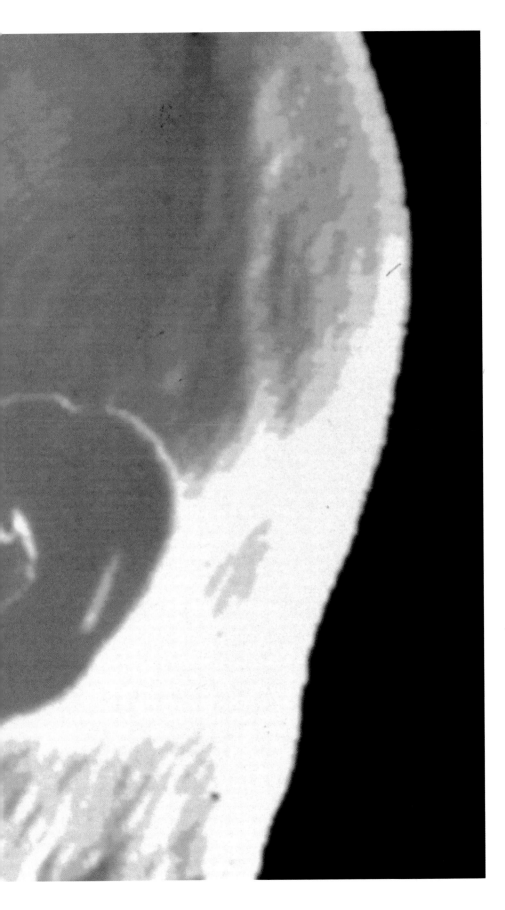

HEALTH care today is a work of both science and art. Sophisticated imaging systems see structures that eyes alone can never see; ultrasound hears a rush of blood that ears cannot. Satellites beam information to doctors who are far removed from the surgical scene, and robots wield the surgeons' tools. Medicines are designed by computers and delivered by patches, sprays, and implants. Surgery is performed through tiny incisions. Re-engineered genes are piggybacked onto viruses and steered through a patient's body to repair a defect or replenish a deficiency. Radiation pellets and rays treat cancers. Gerontologists are even exploring ways to tinker with the genetic clock, their aim being to slow it a bit. If they succeed, and if we can eradicate major diseases, we will live out our lives with perhaps a score or more years added, and we will do so in good health.

Magnetic resonance imaging (MRI) uses radio waves to reveal the brain's speech center.

SEE ALSO

Computers • 236

Fast Track • 96

Mediums and Messages • 214

Scoping the Body • 198

IF diagnostic equipment has a venerable grandfather, it is the x-ray machine. Still the quickest way for us to take off our skins and pose in our bones, as someone said many years ago, it images our skeletons and internal organs, using controlled beams of highly energetic electromagnetic rays discovered in 1895 by Wilhelm Conrad Roentgen, a German physicist.

X-rays pass through flesh and thick paper but are stopped and reflected by bones and metal, a characteristic that—when the rays are on their way to a negative photographic plate—results in an image of bones, in white. Various shadows and shadings help identify anomalies, such as dense lesions and fluid in the lungs.

Diagnosticians need more refined tools, though, because x-rays can miss many structures and abnormalities, and they project only two dimensions. Computerized axial tomography, called the CT scan, is one of many. Hailed as the greatest advance in radiology since Roentgen's discovery, it emerged in 1972 largely through the research of Dr. Allan MacLeod Cormack. A South African native, Dr. Cormack was a professor of physics at Tufts University and shared a Nobel Prize for the rationale behind CT. Linking x-ray and digital technology, a CT scan shows the body in cross sections, from which three-dimensional images are constructed. An x-ray tube revolves about a patient's head, for example, converting the images into a digital code. Differences in density between normal and abnormal tissues are revealed, as well as bone details and the location of tumors and other signs of disease.

A step beyond the CT is magnetic resonance imaging (MRI), which also produces cross-sectional pictures. While a patient lies in a tunnel-like chamber, surrounded by electromagnets creating an intense field, the MRI unit reads the radio signals that return from hydrogen atoms in the water molecules of bodily tissues. These signals are fed into a computer and converted into detailed images of soft tissues, such as those in the brain and spinal cord.

Diagnostic tool (below): A CT chamber focuses x-rays to create a sectional image of the human body's internal structures. As tissue absorbs radiation, computers calculate the amount absorbed at each point scanned and then translate the data into a three-dimensional picture on a viewing screen.

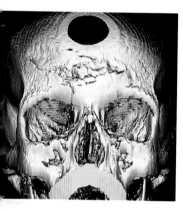

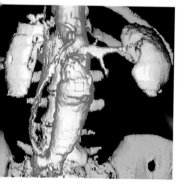

Radiography: An x-ray of a human skull (top) becomes a 3-D image when the technique combines with computer technology in a CT scan.

Seen in a CT scan (above), a potentially dangerous aneurysm distends a major abdominal artery (in red) supplying blood to the lower body.

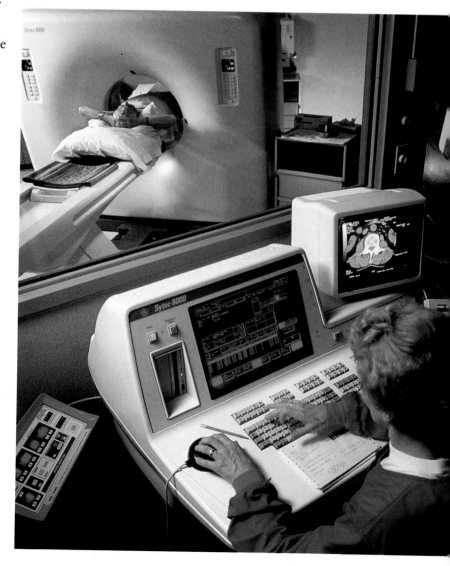

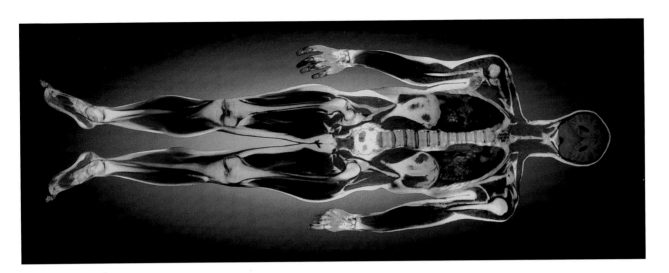

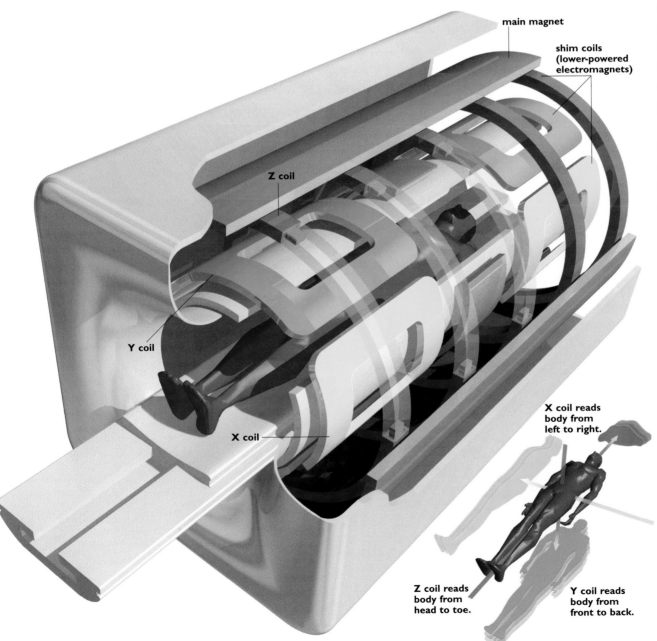

main magnet

shim coils (lower-powered electromagnets)

Z coil

Y coil

X coil

X coil reads body from left to right.

Z coil reads body from head to toe.

Y coil reads body from front to back.

Laid bare by an MRI, a body (above) resembles a model in a medical exhibit, not only because of its appearance but also because of the vital information it provides.

MRI (left): Electromagnetic coils in an MRI unit scan a body with radio waves, exciting hydrogen atoms in bodily tissues. The imaging device then reads signals returned by the atoms to create cross-sectional, 3-D views. Low-powered magnets, called shim coils, control the main magnetic field to vary field strength, setting up gradients in different planes. The red coils "read" from head to toe; green, from front to back; and yellow, from left to right. In this way, each portion of the body can be identified with magnetic coordinates and rendered into a computerized, sectional image.

SCOPING THE BODY

SEE ALSO

Computers • 236

Laser • 206

Supersonic • 116

X-Rays, CT, MRI • 196

IN the classic science-fiction movie, *Fantastic Voyage,* a group of scientists and their submarine are miniaturized and injected into a patient, where they get a first-hand look at organ systems and the interaction of defending antibodies and invader-antigens. Endoscopy affords a somewhat similar view. With flexible fiber-optic tubes inserted through such natural openings as the nose, mouth, anus, urethra, and vagina, doctors can examine internal organs and other structures. One form of endoscopy—a colonoscopy—lets a diagnostician search the large intestine for bowel diseases and bleeding. Another approach involves making a small incision in the abdominal wall and inserting a narrower lighted instrument called a laparoscope. Special instruments may be attached to such tubes, allowing surgeons to remove lesions and collect samples for biopsy.

Because the procedure is uncomfortable and has a small risk of perforation and infection, doctors have sought a noninvasive way to get a direct look. Enter virtual endoscopy, a blending of CT and MRI scans with high-performance computing. The scans can provide simulated visualizations of specific organs in 3-D animated form. A virtual endoscopist, seated at a workstation, views the inner anatomy while manipulating a computer mouse on a "flight path" through the body. Virtual endoscopy lets diagnosticians change direction, their angle of view, and the scale; they can also shift immediately to new views.

Surgeons (below left) examine a gallbladder by viewing its image on a screen. With endoscopy, doctors can peer inside a body by looking through flexible, light-carrying tubes—not into large incisions.

Keyhole surgery (below): Wielding a laser instead of a scalpel, a surgeon performs a laparoscopic operation through a narrow, lighted tube inserted into a tiny incision. Doctors can also collect biopsy samples by using instrument channels.

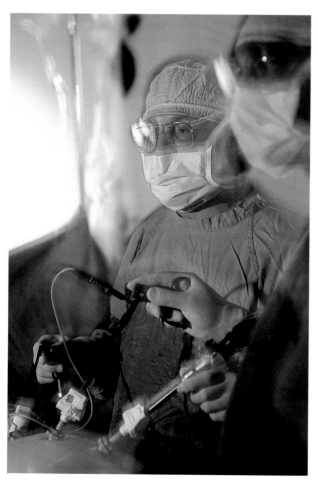

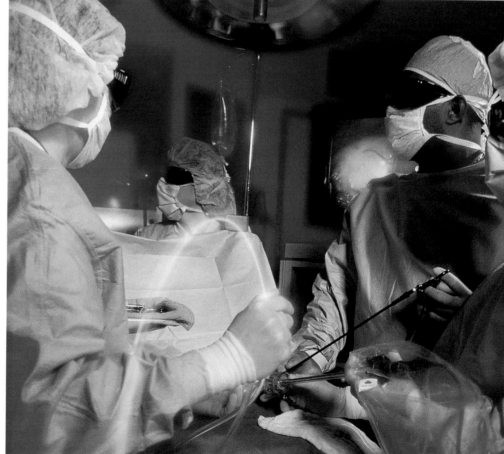

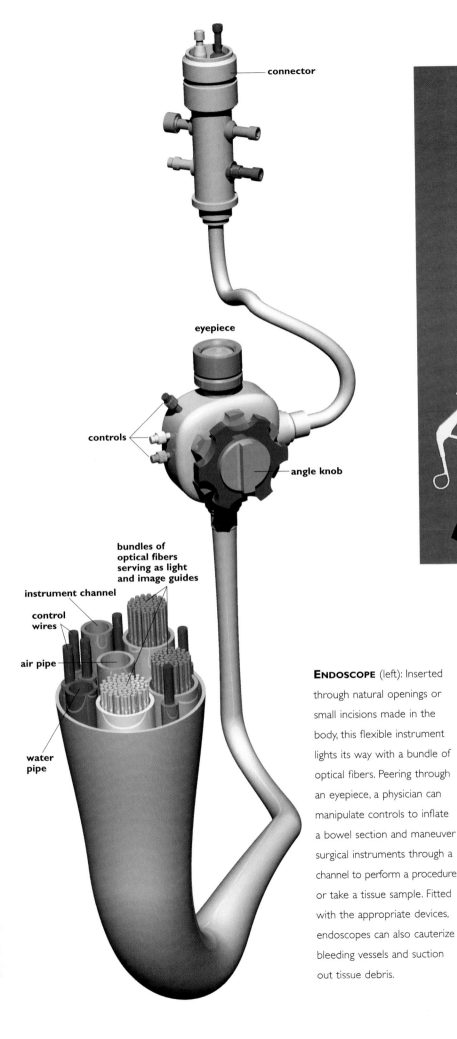

connector

eyepiece

controls

angle knob

bundles of
optical fibers
serving as light
and image guides

instrument channel

control
wires

air pipe

water
pipe

ENDOSCOPE (left): Inserted through natural openings or small incisions made in the body, this flexible instrument lights its way with a bundle of optical fibers. Peering through an eyepiece, a physician can manipulate controls to inflate a bowel section and maneuver surgical instruments through a channel to perform a procedure or take a tissue sample. Fitted with the appropriate devices, endoscopes can also cauterize bleeding vessels and suction out tissue debris.

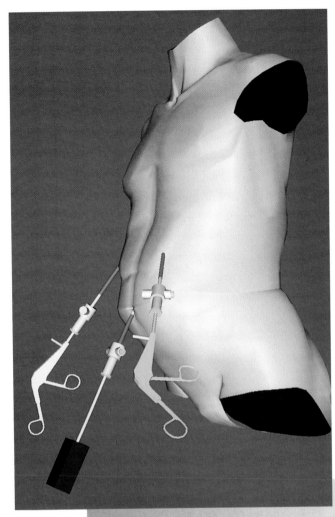

MINIMALLY INVASIVE SURGERY

Increasingly, surgeons are operating through small slits rather than large incisions to remove ovarian cysts and herniated lumbar disks, do hysterectomies and bowel resections, or repair hernias and torn knee ligaments. A gallbladder operation, which once required a five- to eight-inch incision and a recovery period of a month or two, now can be done on an outpatient basis, with a two-week recovery. Doctors make three tiny incisions in the abdomen and insert surgical instruments and a small video camera. They separate the gallbladder from the liver and other structures and remove it through one of the incisions.

SEE ALSO

Drugs • 208

Spare Parts • 202

NESTLED in the rear of the abdominal cavity, the two bean-shaped kidneys function primarily as part of the body's waste-management system, using urine to clear it of urea, the principal breakdown product of proteins and blood plasma salts. Controlled by hormone action, kidneys also maintain water balance and regulate the body's acid-base balance. A person can function with one kidney, but urine production halts and waste products accumulate in the blood when an inflammation or infection causes both organs to fail. Water also accumulates, and chemical concentrations ordinarily regulated by the kidneys are thrown off balance. In advanced form, kidney failure often requires a transplant or a cleansing process known as dialysis.

Peritoneal dialysis takes advantage of the semipermeability of the peritoneum, the delicate membrane lining the abdominal cavity. After an incision is made in the abdominal wall, irrigation fluid is pumped in from a thin plastic tube; waste products enter the solution from the blood and are withdrawn. In hemodialysis, blood from an arm vein is pumped through a tube to a machine that uses an artificial membrane to filter it; the cleansed blood is kept warm and returned to the body in the same vein.

The internal structure of a pair of healthy kidneys (above) and the shape of the liver—the large dark area above the left kidney—emerge when scanned by an MRI unit.

DIALYSIS: A kidney dialysis machine (left) filters waste products from the blood of patients who have kidney disease; in other cases, it removes poisons and drugs. The machine allows wastes, but not blood cells, to pass through a semipermeable membrane. Blood flows from a patient's vein through a tube of semipermeable membrane in the machine's tank, which holds a special solution known as dialysate. Through selective diffusion, unwanted materials in the blood pass through the membrane; the machine pumps them out, while the cleansed blood returns to the patient's vein.

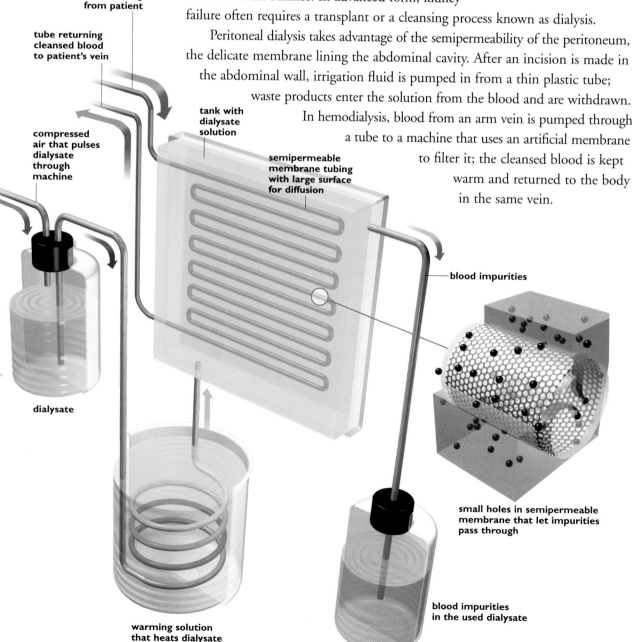

tube coming from patient

tube returning cleansed blood to patient's vein

compressed air that pulses dialysate through machine

tank with dialysate solution

semipermeable membrane tubing with large surface for diffusion

dialysate

blood impurities

small holes in semipermeable membrane that let impurities pass through

blood impurities in the used dialysate

warming solution that heats dialysate

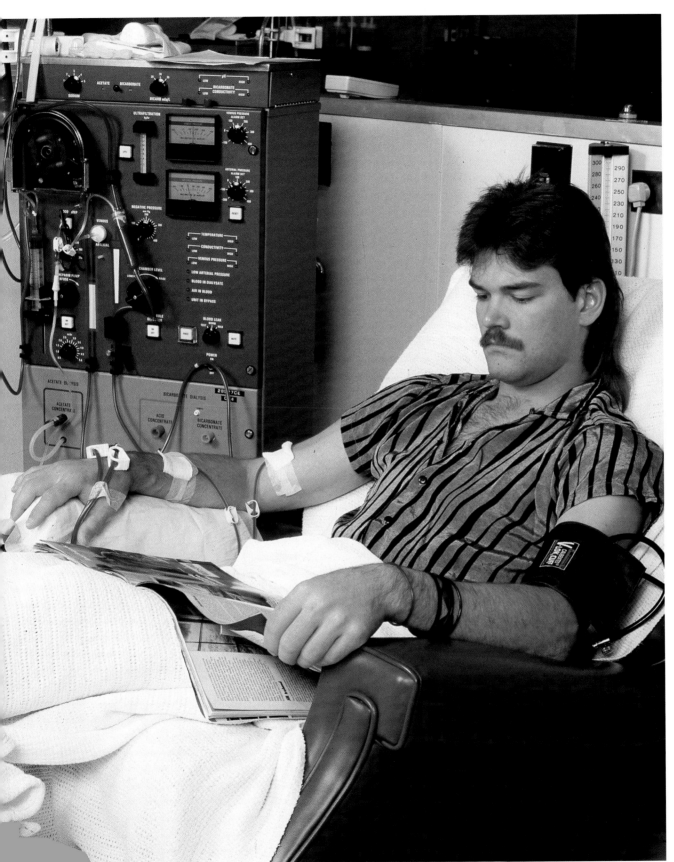

"Artificial kidney":
Conceived of in 1913 as a short-term treatment for reversible acute renal failure, dialysis machines now help transplant patients get by as their bodies accept or reject kidney grafts. In hemodialysis, a tank in the machine holds the dialysate fluid, and a semipermeable membrane filters impurities from the blood. In peritoneal dialysis, the patient's abdominal cavity holds the fluid, and the peritoneal membrane lining the cavity filters out waste products. Either the machine or the patient can exchange used fluid for fresh dialysate.

SPARE PARTS: IMPLANTS FOR THE BODY

SEE ALSO

Faux Fabric
• 138

Medical Automation
• 204

MODERN technology deserves enormous credit for giving patients artificial heart valves, implantable cardiac defibrillators, titanium and ceramic hip joints, throat implants made of synthetic bone to improve swallowing and speaking, and the numerous other spare parts people need these days. But we have understood for centuries that many of our bodily parts move and work like mechanical structures and undergo the same stresses and strains, so it should be no surprise that the use of artificial replacements for what is missing or defective goes way back. Dental implants were used by the ancient Egyptians and by Central and South American cultures, and artificial limbs may have been known to early Greeks and in old India.

Suitable materials are the key to a successful implant. Because the body rejects both organic and inorganic intruders, bioengineers try to design materials that make the interaction more lasting and trouble free. Aluminum compounds are used in dental and orthopedic prostheses because their interaction with surrounding tissue is minimal, and they have low levels of friction and wear. Synthetic polymers that resist water and oxidation and can also insulate and lubricate are used to protect electrical implants such as cardiac pacemakers. A number of porous materials and absorbable composites allow for natural tissue growth in and around some implanted devices, and special coatings on metal implants reduce corrosion. A coating of genetically engineered cells may one day prevent an implant from being rejected or from being choked by scar tissue the body makes when invaded.

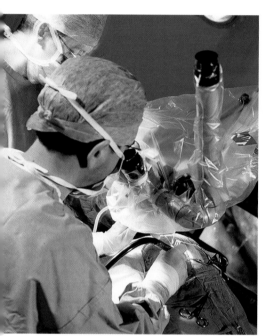

Ear work (left): Eyes glued to a microscope, a surgeon proceeds carefully, inserting a cochlear implant—complete with electrodes—deep within a patient's inner ear.

Racing the clock, a team (below) hurries an organ to a patient waiting for a transplant. A medical team usually keeps a donated heart in iced salt water; a technique that allows organs to remain in optimal condition for three to five hours after removal.

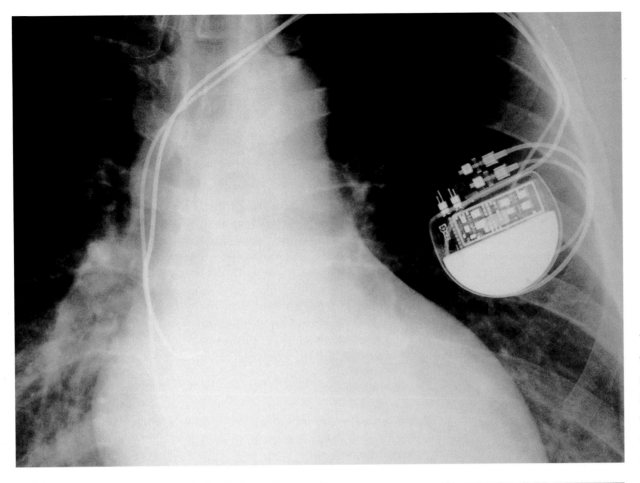

The beat goes on under the influence of a battery-powered pacemaker (left). Implanted in the chest cavity, the device signals contraction of the heart muscle and maintains a normal rhythm.

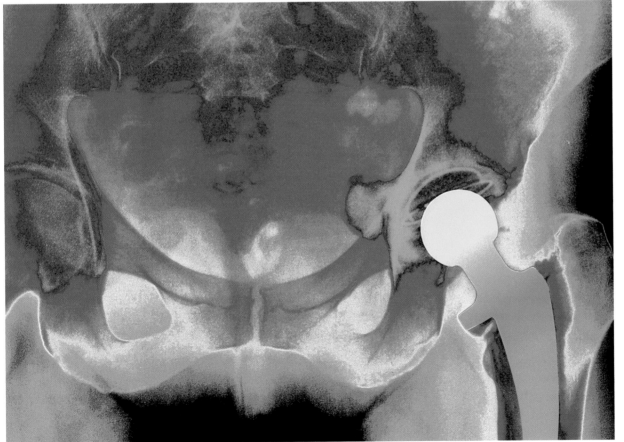

Hip joints (left): Like the ball-and-socket joint found in automobile steering mechanisms, the counterpart in a human hip—ball at the end of a thighbone, rounded socket in the pelvis—permits axial movements. When a prosthetic hip joint replaces a person's failing one, the human joint becomes even more like the mechanical version. As shown at left, a metal ball (in yellow) fits into the natural socket, while its metal shaft (in green and blue) fits into the upper part of the thighbone.

MEDICAL AUTOMATION

SEE ALSO

**Assembly
Line • 84**

**Scoping the
Body • 198**

**Spare Parts
• 202**

PEOPLE once thought of robots only as amazing creatures in science fiction or as powerful automatons working on industrial assembly lines. No longer. In today's world, automated machines also play important roles in health care: Robotics and other forms of computer-assisted medical intervention make surgery more precise, improve a surgeon's dexterity, reduce complications, and help lower the costs of medical care.

Medical robots are generally assumed to be human-like in appearance, but most are not. More often, they are fairly shapeless, computer-controlled machine tools. Robots hold and guide laparoscopic cameras or other delicate instruments during surgery, and they perform such intricate tasks as automated precision-milling to ensure a perfect fit of bone and implant during, say, hip-replacement surgery. Some finely tuned robots provide what bioengineers call telepresence—the feeling an operator has when moving scissors-like devices connected to a robot's arms, a sense that the robot's eyes and hands are actually extensions of the operator's own eyes and hands. One robotic arm, ten inches long and developed at the Jet Propulsion Laboratory in Pasadena, California, is agile, steady, and so precise that it can move surgical tools a mere 800 millionths of an inch, an advantage when performing delicate microsurgery in a brain or an eye. Other robotic systems extend human manipulation capabilities. They position and guide needles to do brain biopsies, and they ensure that surgical instruments are in the right place in the nasal cavity during a sinus operation. They help place radiation pellets into diseased organs, and they accurately insert screws during spine surgery.

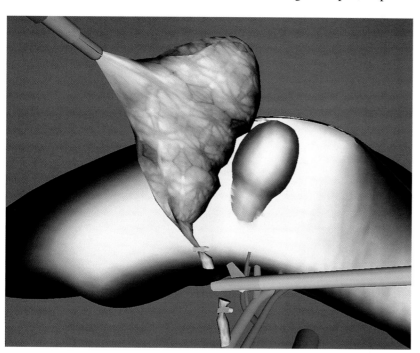

SURGICAL SIMULATION
(above): During a virtual reality operation on a gallbladder, digital data and computers manipulate the surgical instruments and the target organ.

Lending an arm (right): A computer-generated image allows a surgeon to guide computer-controlled instruments in an intricate operation on the brain.

Robots have nonsurgical medical uses as well. "Patient" robots designed to look like people have been used to train medical students. One classic Japanese model even has vital signs—blood pressure, respirations, and heartbeats—that are programmed into its computer. When a student applies a stethoscope to the robot's chest, the heartbeat and erratic rhythms can be detected. Fingers pressed against the wrist elicit a pulse, and a light shined into the robot's eyes alters the pupils. The robot can also be revived after its heart and respirations are stopped: When artificial respiration is done correctly, the heart resumes beating, blood pressure rises, and the robot's pale lips show color.

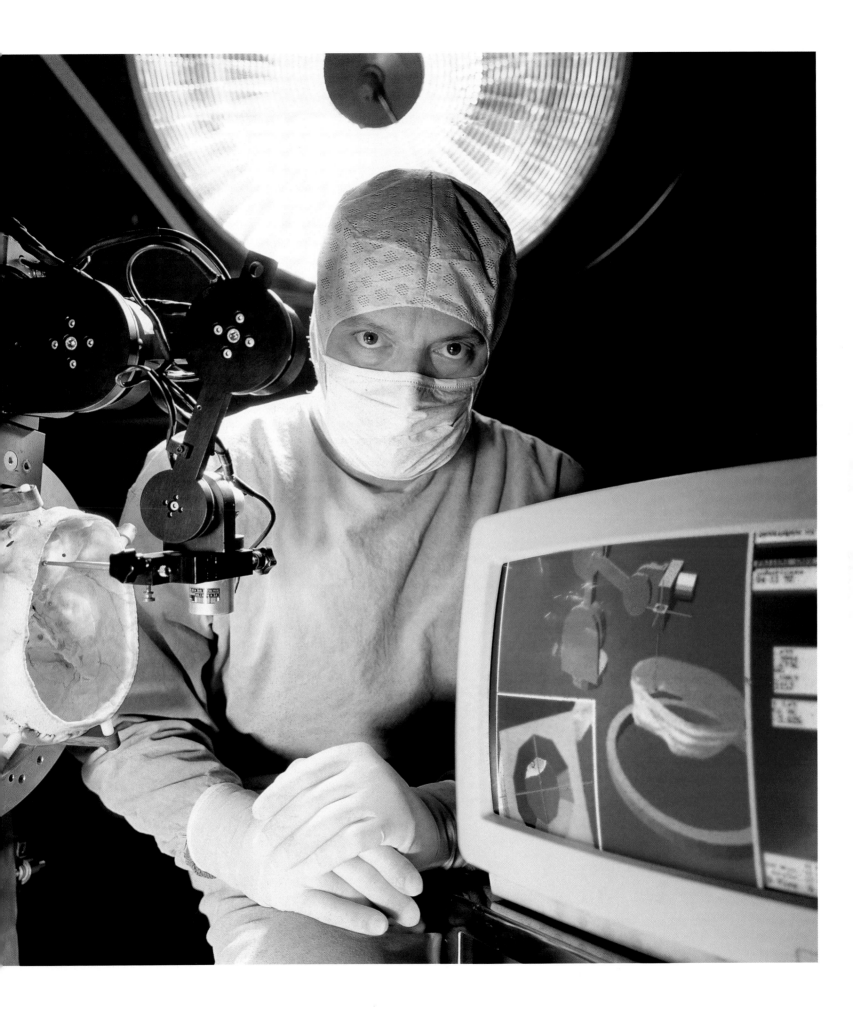

THE LASER

SEE ALSO

Bar Codes • 232

Lasers and Ink • 228

Musical Numbers • 152

Telephone • 212

During eye surgery at Johns Hopkins University Hospital (below), a trail of light marks the movements of a surgeon's hand. Incredibly precise, as well as bloodless and painless, a laser can gently reshape a cornea, repair a retina, and lift an eyelid, all without damaging surrounding tissue.

LASERS carry thousands of telephone calls simultaneously over fiber-optic cables. They play CDs, slice steel beams, guide missiles, measure distances, fashion suits and semiconductor chips, and bore holes in coins or diamonds in a fraction of a second. They also help heal. Medical lasers can vaporize brain tumors, repair detached retinas, spot-weld tissue grafts, cauterize the lining of the uterus to stop prolonged bleeding, and clear blocked fallopian tubes. In addition, laser cosmetic surgery can obliterate tattoos and birthmarks and smooth facial wrinkles.

Generally referred to as bloodless and knifeless surgery, laser treatment seals blood vessels as it cuts; it sterilizes at the same time. Its precision makes it ideal for microsurgical procedures, such as eye repair, reattaching retinas, controlling retinal deterioration in diabetics, and removing cataracts.

Laser stands for "light amplification by stimulated emission of radiation," but its light differs from that of the sun or a bulb, which radiates in every direction. Laser beams are concentrated and narrow, and they move in the same direction. Moreover, they can focus intense heat into small spots: At a width tinier than a pinpoint, a laser can generate a temperature of 10,000°F.

Wavelength and power are determined by a laser's makeup. The solid-state laser, consisting of a rod of ruby crystal coated with a reflecting film, generates bursts of light over brief periods. Argon gas is used as a laser medium in eye surgery because the beam's energy can reach the retina without being absorbed by eye fluid. The carbon-dioxide laser used in gynecology has a beam that is absorbed by substances containing water. Human tissue is about 80 percent water, so the laser can vaporize the diseased cells in its target zone.

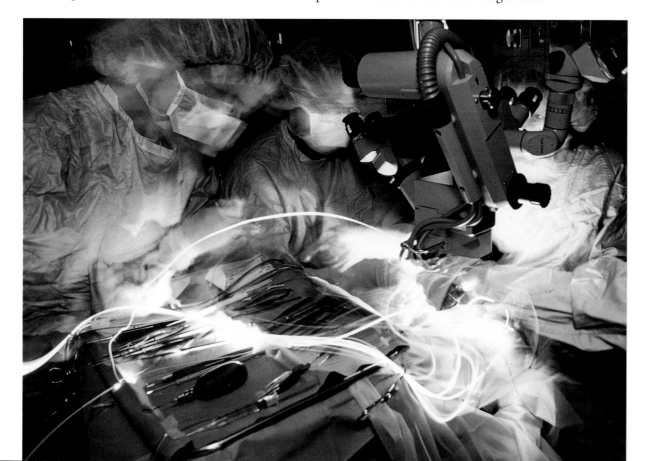

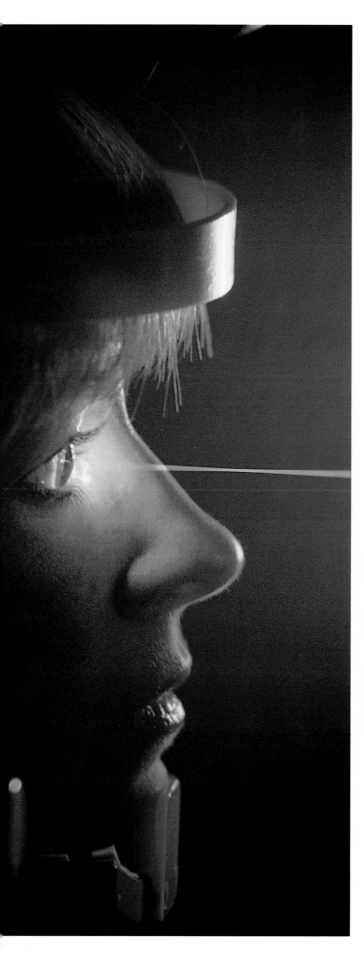

Not batting an eye, a patient (left) stays still during laser surgery. A high-intensity beam of laser energy focuses with pinpoint accuracy while operating in one of the human body's most delicate areas.

Preliminary diagnosis (below): This laser works much like a bar code scanner. The variations in light reflected from diseased and normal tissues help doctors detect cancers and other abnormalities early.

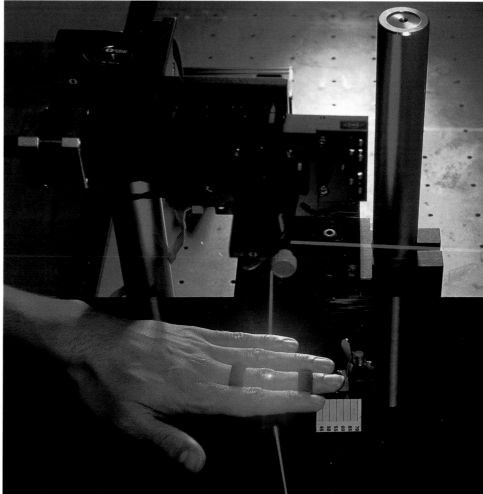

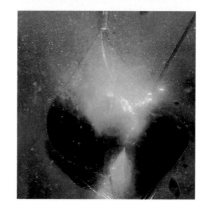

Laser lithotripsy (left): In an experimental procedure, pulses of laser light create stone-smashing shock waves that might eradicate gallstones in an endoscopic basket. Often used to fragment kidney stones, lithotripsy generally transmits ultrasound shock waves through water.

DESIGNING AND DELIVERING DRUGS

SEE ALSO

Computers • 236

Kidney Dialysis • 200

Wonder Fabrics • 140

CANADIAN physician and educator Sir William Osler had his finger on the pulse of hypochondriacs and quick-fixers when he said that a desire to take medicine is perhaps "the great feature which distinguishes man from other animals."

In just one year, pharmacists in the United States fill some two billion prescriptions. The country's drug companies pour 24 billion dollars annually into research and development, spending a total of 500 million dollars, on average, for each new drug that reaches the market. Technology gets all those drugs off the drawing boards and into human (and animal) bodies, and the design and delivery of today's medicines is a paradigm of what scientists from many branches can accomplish in concert. Traditionally, creating a drug involved testing many compounds, adding them one at a time to cell cultures and enzymes to determine which had an effect, and then laboriously testing those to see if they needed to be modified. Computers eliminated some of the drudgery by simulating, for example, protein receptor sites on a virus or on a disease-associated enzyme and then providing a model of a potentially useful drug molecule that might fit the receptor. If it fit, a drug would be designed that fooled the receptor into binding with it.

Taking drugs once meant that a patient had to either swallow them or receive an injection. But both methods have drawbacks. Swallowed drugs may not cross the intestinal lining and enter the bloodstream, or they may be broken down too quickly to have an effect. Injections can be expensive and difficult to administer correctly. Newer delivery methods can compensate: Drug-impregnated, foil-backed skin patches bypass the digestive system and let medication be drawn into the skin via a tiny electric current; nasal sprays take advantage of the nose's permeable mucous membrane, which acts as a portal into the circulatory system. New horizons have been opened by biodegradable drug implants, insulin pumps, and time-release medications.

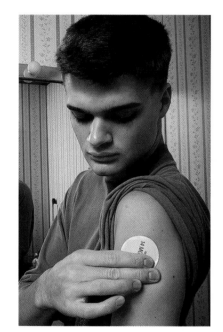

Applying a patch, a young man tries to kick the tobacco habit (above). Patches release nicotine through the skin to lessen withdrawal symptoms smokers may feel when they quit; by dispensing nicotine in decreasing doses, patches can help smokers wean themselves from tobacco. Small molecule drugs, such as nicotine, pass most easily through human skin, but patches may soon dispense insulin and other large-molecule drugs as well.

FINDING THE KEY (below): A lock-and-key mechanism relying on cell receptors helps drugs work within the cells of the body. In one variation (at left), a drug mimics and reinforces natural messenger molecules that the body produces to call up disease-fighting white blood cells. In another variation (at right), a drug blocks the receptor sites to keep out molecular messages and allow the cell to function normally. Pharmacologists can design drugs with molecules that match protein receptor sites on viruses or disease-related enzymes, fooling the sites into accepting the new drugs.

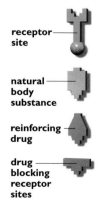

receptor site

natural body substance

reinforcing drug

drug blocking receptor sites

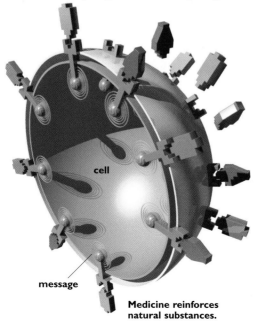

message

Medicine reinforces natural substances.

cell

Medicine blocks natural substances.

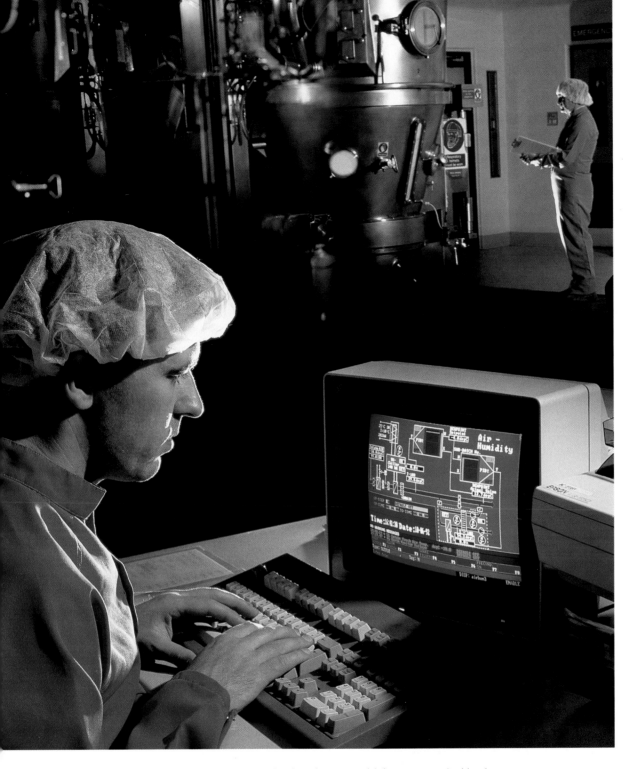

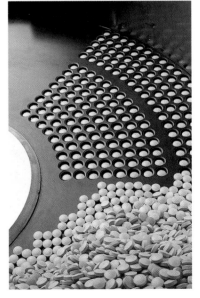

"Fire burn and cauldron bubble," intoned the witches in *Macbeth*. No cauldrons sit in the control room of a modern drug manufacturing plant (above), but reaction vessels abound. Drugmakers rely on computers as well, not only to model and test the "fit" of new drug molecules and receptor sites but also to monitor the environmental conditions in a plant. Here, humidity data essential to quality control in the mass production of pharmaceuticals appear on a technician's screen. Automatically and meticulously, pharmaceutical equipment fills **vials** (above right) to programmed levels. Resembling pieces for an ancient board game, the **little pink pills** (right) that a doctor will prescribe get sorted and readied for inspection. Even though they compete with sprays, patches, and implants, pills remain the most effective way to dispense medication.

COMMUNICATION

WE communicate information in many ways, passing it along through our speech and writing, symbols and signs. We also use visual images, gestures, facial expressions, and bodily attitude as we exchange and share facts and notions. Animals, too, have many ways of communicating with one another, and plants may have a system of communication as well. But as far we know, only we humans have developed the ability to reconstruct and vastly improve our ways of communicating. To do so, we tame and bridle such natural forces as electricity, radio waves, light, and sound, using them alone or in combination. From the cellular telephone to e-mail, from telecommunication devices for the deaf to wearable computers, science and technology have left an indelible imprint on how we get our messages across.

Never out of touch, even on a lake, a kayaker phones home while others wait their turn.

THE TELEPHONE

SEE ALSO

Mediums and Messages • 214

On Call • 218

Radio • 148

Sending Signals • 150

Sky Telephone • 216

ALEXANDER Graham Bell's first telephone bore little resemblance to the sleek cordless telephones we use today. Instead, it was a rather odd-looking "gallows frame" rig consisting of vibrating metal diaphragms and a bar magnet wrapped in wire.

The principles behind the phone's function are essentially unchanged, but the way the device processes and transmits voices is now vastly different. Basically, a telephone is a mouthpiece with a tiny microphone and an earpiece with a receiver. When a person speaks into the mouthpiece, sound waves cause the microphone's diaphragm to vibrate. Speech is converted to electrical signals that can be transmitted over cable, radio, and microwave connections, or to pulses of laser light carried over fiber-optic equipment. Automatic switching equipment identifies the number dialed and makes the connection. For years, the electrical signals in vocal transmissions were analogs—they were analogous to the vibrations in conversation—but modern systems convert the sound waves into digital information at the local exchange, hiding it in a binary code of 1s and 0s. On the receiving end, the process is reversed by a vibrating diaphragm that picks up the reconverted analog signals and recreates the speaker's voice.

Skeins (left) of multicolored telephone wiring still include simple pairs of copper wire for local calls.

They've got your number: Under the control of a central computer, automated telephone switching equipment (below) identifies a dialed number and sends the call to its destination.

Actress Lili Tomlin (left) portrays "Ernestine," a chatty telephone operator stationed in front of an old-fashioned switchboard. Several decades ago, many women sat in rows of such consoles, plugging in jacks to make connections.

TELEPHONE CALLS (below): In the United States, where 100 million households have telephones, a vast network of some 150 million main telephone lines carries long distance and local phone calls. The process of making and receiving a call involves converting the voice into an analog signal of electrical current and sending it to a local exchange that converts it to a digital code of eight electrical "on" or "off" pulses. Electronic switches combine the binary data from many different conversations and send them over a single wire. The receiving end reassembles the components of each conversation, reconverts them to analog, and sends them to the appropriate phones.

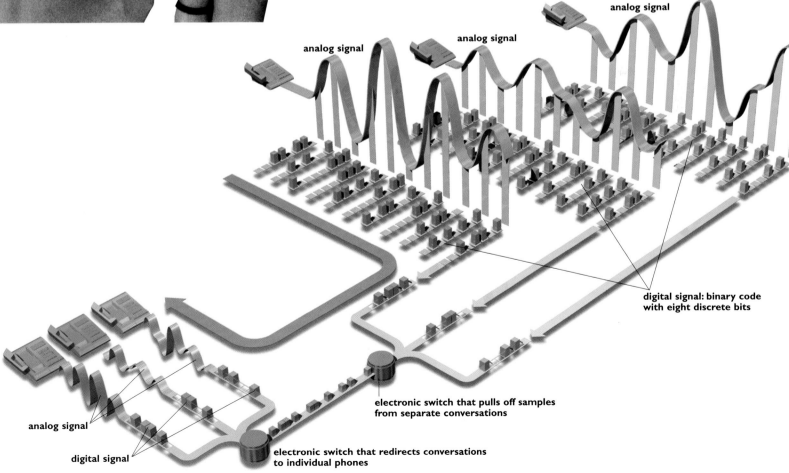

analog signal

analog signal

analog signal

digital signal: binary code with eight discrete bits

electronic switch that pulls off samples from separate conversations

analog signal

digital signal

electronic switch that redirects conversations to individual phones

THE MEDIUMS AND THE MESSAGES

SEE ALSO

Laser • 206

Radio • 148

Sending Signals • 150

Sky Telephone • 216

Telephone • 212

TELECOMMUNICATION—communication at a distance using telephones or radios, for example—relies on electromagnetic waves. Members of this large energy family include radio and light waves, x-rays, microwaves, and infrared and ultraviolet rays. Called electromagnetic because they consist of electric and magnetic fields (which vibrate at right angles to each other), they are high-velocity energy transmitters that move at the speed of light (186,000 miles a second). Their signals and impulses are capable of carrying sounds, words, images, and data. The higher their frequency, the more energy they have; gamma and x-rays are at the high end; microwaves and radio waves are at the low end.

Radio waves exist naturally as radiant energy from the sun, but when they are generated by electricity in an antenna tower, they transport sound and images. In the form of radar, radio waves help detect distant objects and determine their positions; radar operators measure the time the waves take to travel to an object, reflect off it, and return. The visible light segment of the electromagnetic field has also been used to communicate: In laser form it beams through fine strands of pure glass, its digitally coded pulses translated into thousands of telephone conversations that can be handled simultaneously.

Sound also comes in waves, but they are pressure waves, not electromagnetic ones. Best defined as a measurable, mechanical disturbance, sound is amenable to translation into electricity. This is a valuable quality when transmitting voices, whose words, whether flowery or vernacular, must still be reduced to the same workmanlike electrical signals before they can be sent coursing over conventional copper wire or through tubes of glass.

Fiber-optic cables (above): Telephone conversations pulse digitally through cables like these. The glass strands carry tens of thousands of calls at the same time.

WAVELENGTHS (opposite): Electromagnetic radiation ranges from high-frequency gamma rays, at the top of the electromagnetic spectrum, to low-frequency radio waves.

SHEATHED OPTICAL FIBERS (far left), wrapped around a steel strengthening wire, transmit digital data in the form of laser pulses. Each fiber's core and cladding are made of silicon glass and semiconducting materials, and the plastic sheathing prevents stray light from escaping to other fibers. **NARROW-CORE FIBERS** (near left) carry clear signals over long distances, while **WIDE-CORE FIBERS** allow the signal to spread out and blur, limiting data's transmission rate.

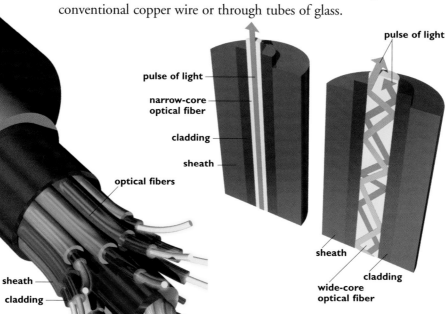

pulse of light

pulse of light

narrow-core optical fiber

cladding

sheath

optical fibers

sheath

cladding

sheath

cladding

wide-core optical fiber

steel wire

core

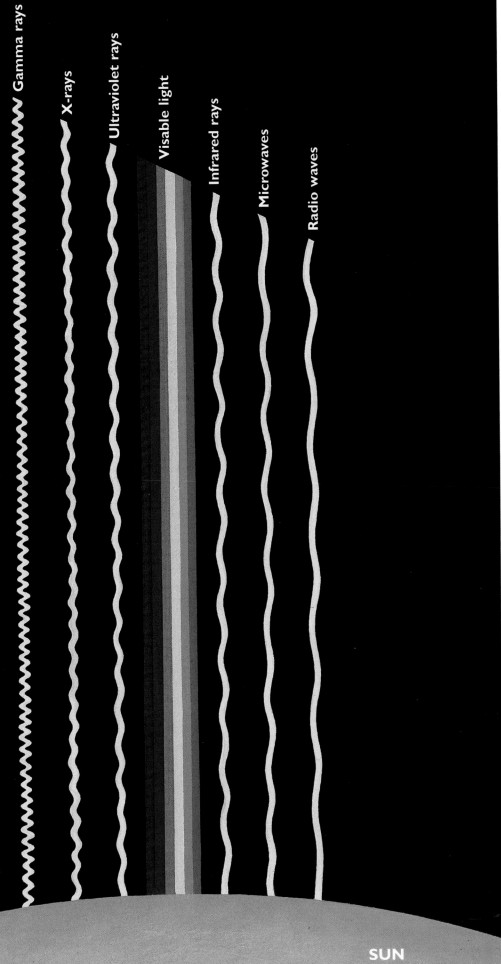

Gamma rays

X-rays

Ultraviolet rays

Visable light

Infrared rays

Microwaves

Radio waves

SUN

THE SKY TELEPHONE

SEE ALSO

Eyes and Ears
• 220

Mediums and Messages • 214

Radio • 148

Sending Signals
• 150

Telephone • 212

THE wedding of the telephone and the radio was perhaps the most significant development in the history of telephony. Roaming from room to room with a cordless phone, or smartly snapping open a compact cellular model while on a stroll in a park, a subscriber to a phone service now has a degree of mobility probably not even imagined by Alexander Graham Bell.

Cellular phones—so-called because they cover compartmentalized, cell-like areas—are more impressive than the now standard cordless ones. They transmit over radio waves via an antenna located in a base station in each cell, or by way of satellites. Currently, the antennas are connected by phone lines to exchanges, which link cell phone users to one another or connect them to parties using conventional phones. Eventually, extensive satellite telephone systems will connect directly with each subscriber's phone, enabling callers to reach out to all corners of the world. Another generation of cell phones will be able to instruct a Global Positioning System satellite—alerted by a 911-like message—to beam a signal to the receiver, thereby determining the exact location of, say, a driver in distress. Similar applications would enable callers to use slot-configured mobile phones as miniature ATMs when they need to download cash from bank accounts to their cash cards.

CELLULAR PHONES (below), with their own radio receivers and transmitters, pick up signals from antennas on base stations located in cell-like areas or from receiver dishes tuned to satellites. The frequencies in a cell group differ from one another, but beyond a group the same frequencies may be reused throughout a network. As a person moves from cell to cell, the call automatically switches to the appropriate frequency. Phone lines connect cell phone base stations to exchanges that handle cellular calls. The exchanges convert cable signals to radio signals routed to other cell phones; they can also switch calls to conventional phones.

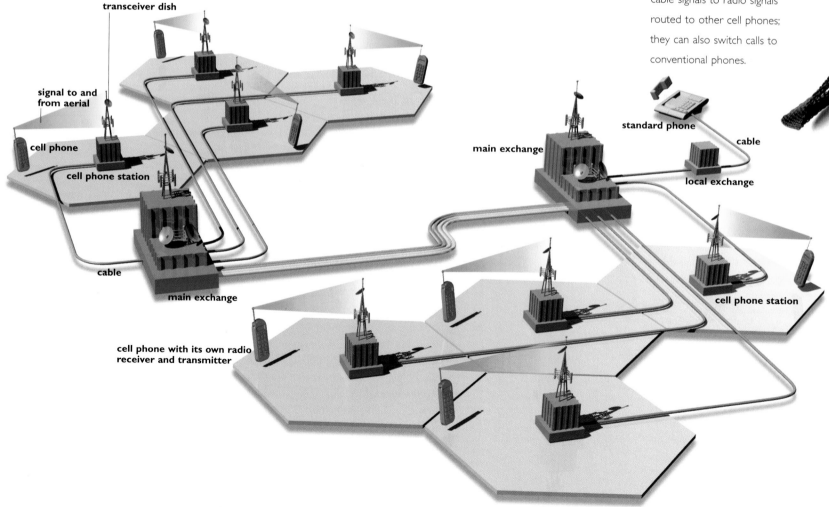

transceiver dish

signal to and from aerial

cell phone

cell phone station

cable

main exchange

cell phone with its own radio receiver and transmitter

main exchange

standard phone

cable

local exchange

cell phone station

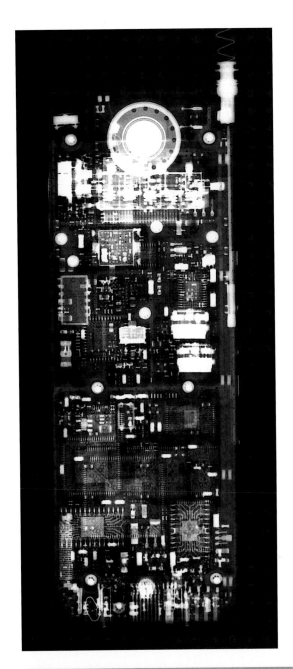

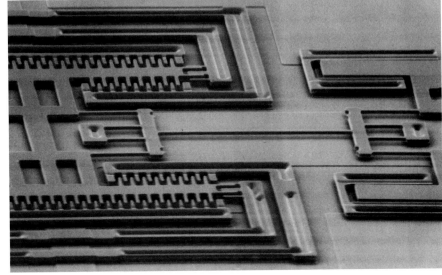

Mobile phone (above): Designed for maximum mobility, a cell phone draws radio signals from the closest antenna in a network. Digital phones transform voices into computer language and prove less prone to unauthorized access than analog phones. An x-rayed **cell phone** (above right) displays its miniaturized components. It relies on some of the same mechanisms used by standard phones—microphone, earpiece, and Touch-Tone dialing—but it has no dial tone, handset, or line cords; it does have electronic helpers, such as signal strength and in-service indicators, built into it. Less than 2/1000 of an inch long, a **micromechanical resonator** (right) oscillates at the same frequency as radio waves that carry a conversation, thus making it possible to extract them from the many signals in the air.

ON CALL

SEE ALSO

Copiers • 226

Mediums and Messages • 214

Radio • 148

Telephone • 212

FAXING (below): When someone feeds a document or a picture into a fax machine, a light-scanner reads the data, and image sensors convert it to electrical pulses. Next, the machine converts the electrical stream to digital form, and a modulator combines it with a carrier wave, allowing the machine to send the information out over an open phone circuit to the recipient's machine. In receiving mode, the demodulator converts the arriving waves into digitized information that prints by the same process employed in a photocopier.

FACSIMILE machines, pagers, and answering machines are the handmaidens of the telephone, and they have become almost as essential to daily life as phones, radios, and electric lights. The facsimile, or fax, is the oldest of the pack, dating back to 1843, when English clockmaker Alexander Bain was given the first patent for facsimile transmission.

Today, faxes are fixtures in homes as well as offices, sending exact copies of documents or pictures around the world in seconds. All it takes is a phone call to connect the recipient and the sender, who feeds the material to be sent into a fax machine. Inside, a light beam plays over the text or photograph and reflects an image of the light and dark portions into an arrangement of photoelectric cells that, in turn, convert the information into electrical current. Amplified, the current is transmitted over the open telephone circuit to the recipient's machine, where a printer reassembles the imagery, line by line, on paper.

Pagers and telephone answering machines take messages for people who are temporarily away from their phones. The answering machine, which is essentially a telephone with a recording device inside, detects an incoming call and, after a preset number of rings, plays a prerecorded message inviting the caller to leave a message. A call made to a pager, or beeper, is converted into a code that can be accessed only by a device that can read it.

Good news travels fast (above): A pager delivers a message that began with a phone call. From a base station, the message traveled via radio waves to a peripheral station that beamed it to the pager.

A modern fax machine (opposite, lower) uses digitizing technology not found in early facsimile machines, which sent news photographs via phone, or in teletypes, typewriter-like machines linked to phone lines that transmitted news letter by letter to receiving machines.

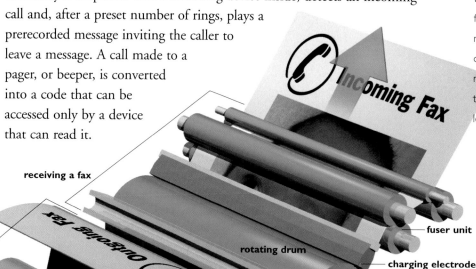

receiving a fax

sending a fax

Incoming Fax

Outgoing Fax

fuser unit

rotating drum

charging electrode

telephone cable

modulator

signal transfer unit

incoming signal

outgoing signal

image sensor

INTERCOM

An intercommunication system, or intercom, provides two-way conversation between rooms. Each position has a microphone and a loudspeaker, generally combined in one unit installed in a wall and connected to a central electric amplifier. Intercom systems also come in other guises. The videophone, for example, was introduced by AT&T at the 1964 World's Fair. While not yet a hot consumer item, buyers can find models with microphones and cameras, and they are amenable to linkup with home computers, television sets, or the Internet.

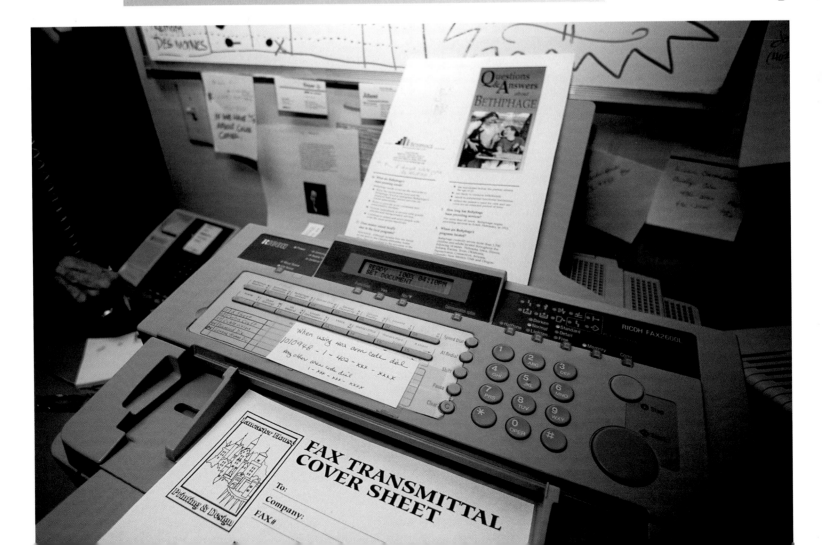

SEE ALSO

ATM • 230

Bar Codes • 232

Radio • 148

Sky Telephone • 216

Television • 156

SPUTNIK 1, the first artificial object to orbit the Earth, was launched in 1957 by the former Soviet Union. Three months later, Sputnik was burned to a crisp, but the era of satellite technology had been launched. Today, communication by radio, television, or telephone is unthinkable without these miniature moons. Carried into orbit by rockets or space shuttles, satellites handle far more than TV sitcoms, news, and phone calls. They also track weather systems, map Earth's features, broadcast navigational signals, and help plan battlefield strategy. Capable of transmitting nearly all kinds of data, satellite radio signals carry up-to-the-moment financial information to Wall Street traders, allow us to swipe our credit cards through machines at gas stations, and provide chain stores with another way to check warehouse inventory.

Satellites can range over the planet, but the so-called geostationary orbiters used for communications and weather observations are synchronized with the Earth's rotation. In an orbit 22,300 miles above the Equator, they take 24 hours to complete a revolution; this "parks" them above the same point on the surface, a position that allows for uninterrupted contact between ground stations in a satellite's line of sight. Three relay satellites uniformly spaced at such a height can easily cover the entire surface of the planet, receiving television and hundreds of thousands of telephone messages from one continent, amplifying them, and relaying them via other satellites in the linkup to other parts of the globe.

CELL DIVISION (below left): Launched into low polar orbits, U.S. Iridium satellites provide voice circuits, data, and paging. Their circular "spot" beams cover "cell" areas some 90 miles across, and they can communicate with each other as well as with Earth stations. Now the mainstay of global communications, solar-powered communications satellites came into use in the 1960s.

Global interconnectivity: A hiker (below) relies on radio data beamed from a Global Positioning System satellite that covers a specific portion of the Earth's surface.

Cells: Circular beams from Iridium satellites are about 90 miles across.

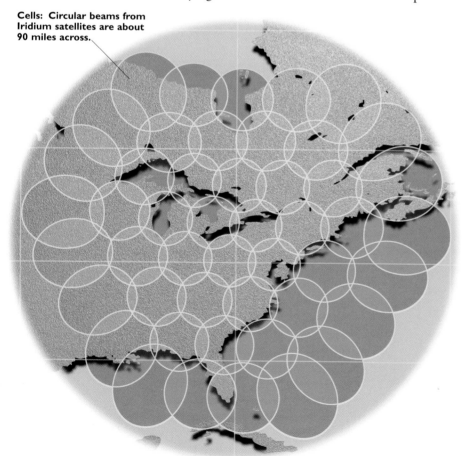

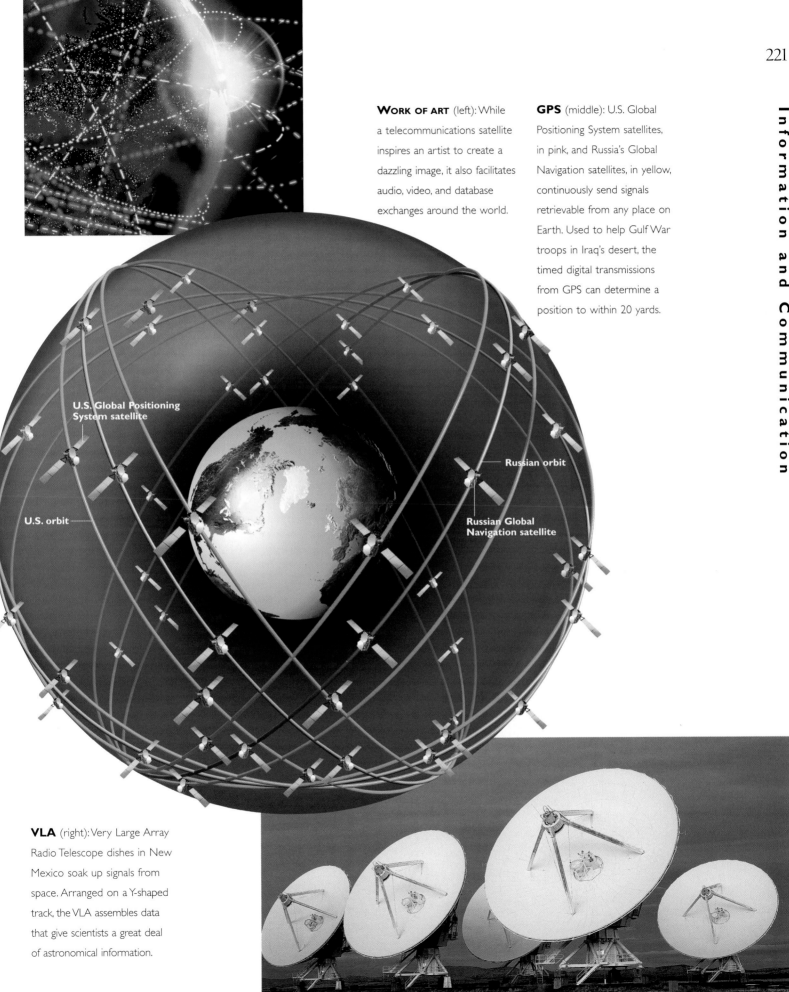

WORK OF ART (left): While a telecommunications satellite inspires an artist to create a dazzling image, it also facilitates audio, video, and database exchanges around the world.

GPS (middle): U.S. Global Positioning System satellites, in pink, and Russia's Global Navigation satellites, in yellow, continuously send signals retrievable from any place on Earth. Used to help Gulf War troops in Iraq's desert, the timed digital transmissions from GPS can determine a position to within 20 yards.

U.S. Global Positioning System satellite

Russian orbit

U.S. orbit

Russian Global Navigation satellite

VLA (right): Very Large Array Radio Telescope dishes in New Mexico soak up signals from space. Arranged on a Y-shaped track, the VLA assembles data that give scientists a great deal of astronomical information.

QWERTY AND BEYOND

SEE ALSO

Computers
• 236

Piano • 146

PATENTED in 1868 by one of its inventors, Christopher Latham Sholes of Wisconsin, the first practical typewriter had keys that worked on a "pianoforte" action: When fingers pressed the keys, a series of connected levers raised the type bars; these struck the paper curved around a roller. In the old models, a ribbon of inked fabric ran across the top from one end of the machine to the other; each type bar had upper- and lowercase letters, and a row of keys held numerals and punctuation marks. Pressing a shift key lowered the type bar, enabling the uppercase letters to strike the ribbon.

The electric typewriter came into use in the 1920s. Early versions each had an electric motor that lifted the type bar and powered the typing stroke. The motor also returned the carriage, turned the roller, and made the keys jump at a typist's touch. The type bar was replaced in some later models by a spherical typing head, an "element" with 88 characters that moved along with an inked ribbon holder as keys were struck. Except for specific controls, a computer's keyboard resembles a typewriter's, keeping the so-called QWERTY configuration, after the first six letters below the numbers. The computer's shift key, however, doesn't lower a type bar. It sends a digital signal to a word-processing program.

Manual typewriter

(below): Typewriters do not write; they only type, as some people say. Still, the machines have churned out prose and poetry for generations of novices and professionals, many of whom could get 70 words a minute and more out of the old manual models.

A **MECHANICAL TYPEWRITER** has type bars linked to key levers. When a typist presses a key (upper right), its type bar strikes paper through an inked ribbon. The **ELECTRIC TYPEWRITER,** which came into use in the 1920s, uses an electric motor to lift the type bar (lower right) and power the stroke. The motor also returns the carriage, turns the roller, and makes keys jump when a typist presses them.

Hands-on experience

(below): Marked gloves help teach the touch method of typewriting, which assigns a specific finger to its own keys. Learning this method enables a person to type without looking at the keyboard.

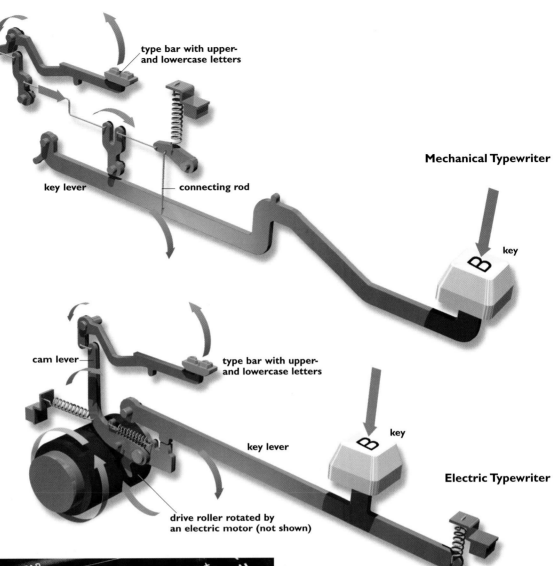

type bar with upper- and lowercase letters

Mechanical Typewriter

key lever

connecting rod

key

cam lever

type bar with upper- and lowercase letters

key

key lever

drive roller rotated by an electric motor (not shown)

Electric Typewriter

IBM SELECTRIC

Instead of type bars, the Selectric typewriter uses a spherical typing head that moves along with an inked ribbon (sample type, top row). Other methods include a "daisy wheel" imprint (center) and dot matrix (below), created by pins striking the ribbon.

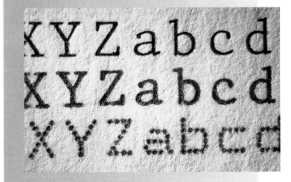

SEE ALSO

Copiers • 226

Lasers and Ink • 228

Presswork (below): A six-color high-speed press rumbles along, blurring the action but not the quality of the print and images it leaves behind on paper.

Gears galore (opposite, top) need tending in an offset printing press, which helps to satisfy the insatiable appetite of businesses for brochures, catalogs, and labels.

UNDISPUTED masters of invention, the ancient Chinese are credited with giving the world some of the earliest examples of printing. Their creations, which took the form of wood-block images on paper and silk, perhaps drew on even earlier techniques developed by the Babylonians and Sumerians to make individual name seals.

The need for mass distribution drives modern printing, and while Gutenberg's invention of printing by movable type, around 1446, revolutionized the way information is disseminated, the advent of so-called web presses and offset lithography put magazines, newspapers, brochures, books, and virtually everything else fit to print into the hands of millions of readers. A newspaper's web press—a machine that prints a continuous roll of paper—can run as fast as 3,000 feet a minute.

Offset lithography is essential to modern printing. This versatile process transfers photographed images and text to a metal plate through photochemical action. The word "offset" means that the ink used to coat the plate does not print directly onto paper. Instead, the inked plate prints on a rubber "blanket" cylinder, and from there images are offset firmly onto paper.

cyan ink trough

oscillating rollers

plate cylinder

yellow ink trough

impression cylinder

blanket cylinder

paper

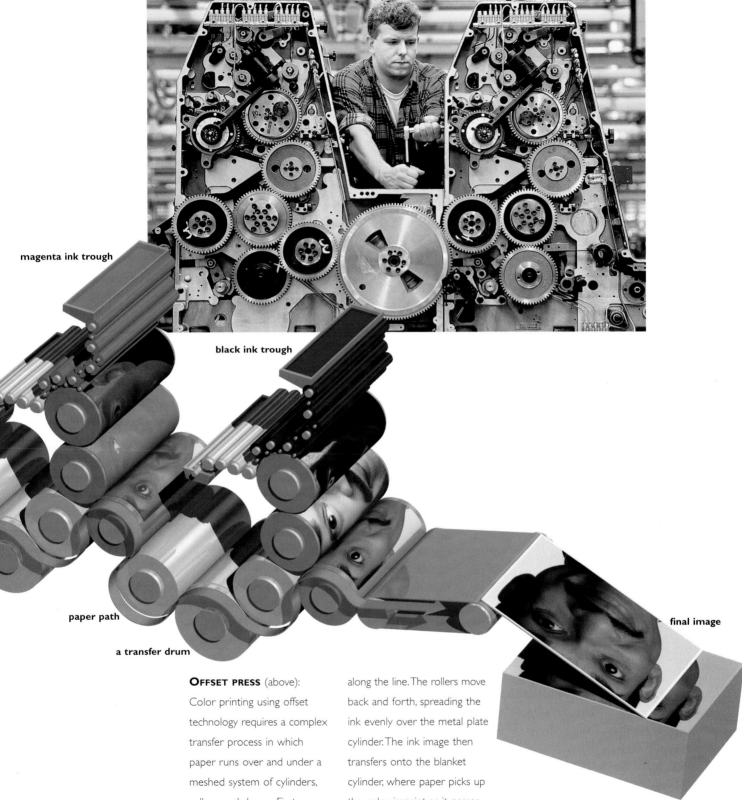

magenta ink trough

black ink trough

paper path

a transfer drum

final image

OFFSET PRESS (above): Color printing using offset technology requires a complex transfer process in which paper runs over and under a meshed system of cylinders, rollers, and drums. First, a photochemical process transfers photographed images and text to a metal plate. Inking rollers apply ink fed from troughs of cyan, yellow, magenta, and black along the line. The rollers move back and forth, spreading the ink evenly over the metal plate cylinder. The ink image then transfers onto the blanket cylinder, where paper picks up the color imprint as it passes between it and the impression cylinder below. Transfer drums move the paper to the next press where the process repeats with another color.

SEE ALSO

Camera
• 160

**Planning
Patterns**
• 136

USING waxy, finger-smudging carbon paper was once the only way to make duplicate copies of printed material. By the late 1800s, this tedious process was replaced by the mimeograph, which relied on stencils cut into coated fiber sheets by a typewriter; a stencil, wrapped around an ink-soaked drum, made copies on the paper as ink seeped through the cut outlines of the characters. Offices also had the hectograph, a contraption loaded with a large roll of gelatin-coated fabric stretched across the top and fixed at the other end. An ink "master" typed on special, coated paper was pressed onto the gelatin sheet and then peeled off, leaving its imprint in a purple, reverse image. To make a copy of the imprint, a sheet of transfer paper was placed on it and run over by a roller.

Today, standard copying is a high-speed process called xerography (from the Greek for "dry writing"). It works on the principle of photoconductivity, by which certain materials conduct electricity under the influence of light. A key element in the copy machine is selenium, a by-product of copper refining and a poor conductor of electricity except when bathed in light. Selenium "reads" the electrical difference between a document's white and dark areas; the white areas, which reflect light, lose their electrical charge; the dark areas do not. Then it converts the pattern into an electrical version of the image. Charged ink powder, called toner, is dusted on; this material is attracted only to the charged image and is then fused onto the paper by heat.

Magic touch (above): An office worker has only to push a button or two to produce quick and fairly inexpensive copies of a document by xerography. Centuries ago, scribes such as learned monks carefully copied manuscripts with quill pens.

XEROX

The Haloid Xerox 914 copier (left), the first dry-process automatic copier on the market, proved a relatively cumbersome apparatus. Introduced in 1959, it made copies on ordinary, untreated paper, but despite problems such as paper scorching, it gave rise to the modern Xerox Corporation and transformed an industry that had several early players. After World War II, 3M and Eastman Kodak introduced Thermo-Fax and verifax copiers. Their duplicates were of poor quality, though, and continued to darken long after they were removed from the machines. Among the most successful products ever made, Xerox copiers were heavily tested: In the 1980s, the company copied more than 20 million pages in one year, just to see if its machines worked.

COPY MACHINE:

Light from a halogen lamp illuminates a document fed into a photocopier (lower). White areas reflect a lot of light, while print reflects little. A mirror and lens send the light onto a revolving drum with a photosensitive coating, which converts the light into a latent electrical image. Positively charged toner adheres to the negatively charged areas on the photosensitive drum that correspond to dark areas on the original document. A charging electrode sends a negative charge to paper on the drum, transferring toner and the translated image onto the sheet. Finally, heated rollers fuse the image to paper.

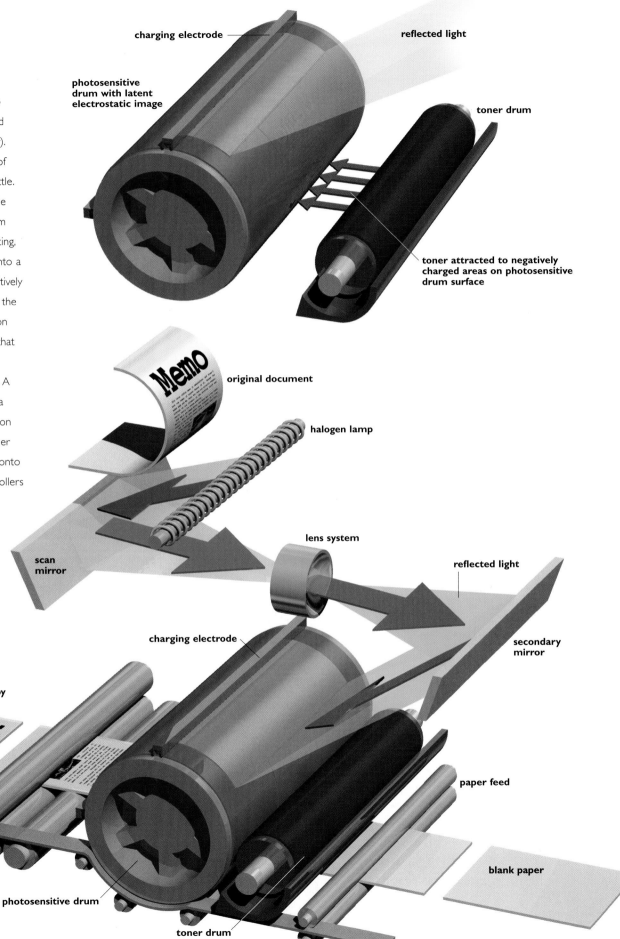

charging electrode

reflected light

photosensitive drum with latent electrostatic image

toner drum

toner attracted to negatively charged areas on photosensitive drum surface

original document

halogen lamp

scan mirror

lens system

reflected light

charging electrode

secondary mirror

final document copy

paper feed

heating rollers

photosensitive drum

toner drum

blank paper

SEE ALSO

Chips and Transistors • 238

Computers • 236

Plastic • 190

ACTIVATED by a plastic card, a bank's automated teller machine (ATM) first determines whether you qualify for service. If you have enough money in your account, you get money from the machine. The key to it all is the dark magnetic strip on the back of the card. Its invisible tracks hold around 200 bytes of information in letters and numbers, and when the card is inserted into an ATM, the data and transaction request are sent electronically to a computer at the financial institution that operates the ATM. There, the account number, the balance, and the user's personal identification number (PIN) are verified. If everything checks out, the computer subtracts the money withdrawn, and a vending machine-like mechanism dispenses the cash, all in a matter of seconds.

Such cards already are being replaced by so-called Smart Cards that do their own computing. Armed with processing power that greatly widens their capability, these cards have electricity-carrying silicon chips that store around 8,000 bytes of information.

ATM (left): The automated teller machine—perhaps the most visible symbol of electronic banking—provides fast cash. The machines not only dispense bills but also take deposits, allow transfers of funds, and answer balance inquiries. To activate an ATM, the user inserts a plastic card with a data-rich magnetic strip into a slot and then punches a personal identification number (PIN) on a keypad. A computer verifies the identification number and cash balance, making note of the transaction requested. When customers withdraw cash, electronically counted money comes out of built-in strongboxes, or cassettes. Many Americans use ATMs about 8 times a month; bank ATMs alone average 6,400 transactions per month.

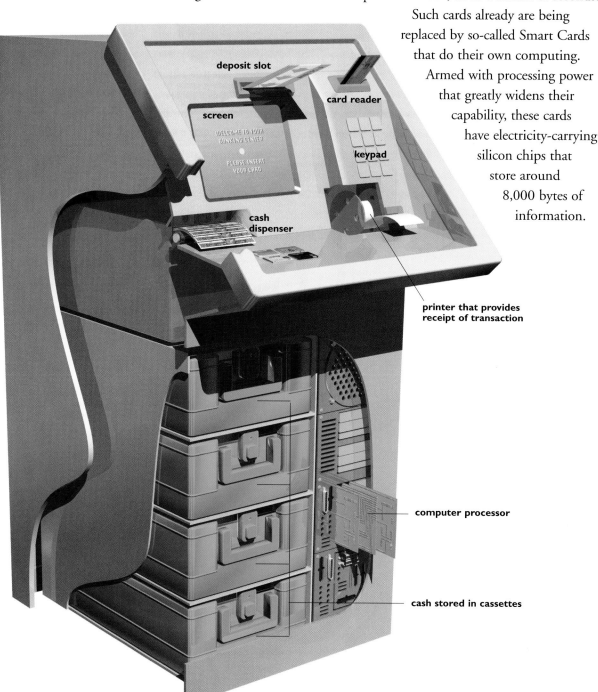

deposit slot

card reader

screen

WELCOME TO YOUR BANKING CENTER

PLEASE INSERT YOUR CARD

keypad

cash dispenser

printer that provides receipt of transaction

computer processor

cash stored in cassettes

Mr. Plastic Fantastic:

Walter Cavanagh makes
quite a fashion statement
with his plastic-lined coat.
His numerous cards speak
to the billions of plastic
transactions that occur each
year in the United States.

BAR CODES AND SCANNERS

SEE ALSO

Computers
• 236

Laser • 206

On Call
• 218

USING computer technology's omnipresent servant, the binary system, bar codes represent the decimal price and other details about an item in a series of parallel vertical lines and white spaces. Such codes also are used to track documents and packages, gene sequences in databases, and books in libraries and bookstores, and they are standardized. For example, bar codes at supermarkets and retail stores come under the Uniform Product Code, which assigns a unique numerical signature to grocery products; the UPC identifies the manufacturer and the product. Bookstores use codes set by the International Standard Book Number (ISBN) system; they identify a book's price and the country that published it.

When a cashier passes a handheld scanner over a bar code or runs an item over a built-in scanner in a supermarket checkout lane, a laser beam set to a specific frequency reads the binary code in the bars, identifying the item and its price. The information is sent to the store's main computer for processing and then is flashed onto the cash-register display.

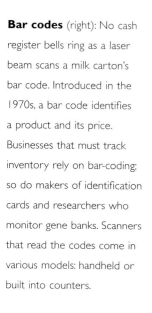

Bar codes (right): No cash register bells ring as a laser beam scans a milk carton's bar code. Introduced in the 1970s, a bar code identifies a product and its price. Businesses that must track inventory rely on bar-coding; so do makers of identification cards and researchers who monitor gene banks. Scanners that read the codes come in various models: handheld or built into counters.

Gatekeeper (right): Making a swipe at a Hong Kong turnstile, a commuter uses a fare card to open a gate that "reads" and then responds to a coded signal in the card's magnetic strip. A convenient alternative to pockets full of coins and tokens, such cards also keep a record of the value remaining.

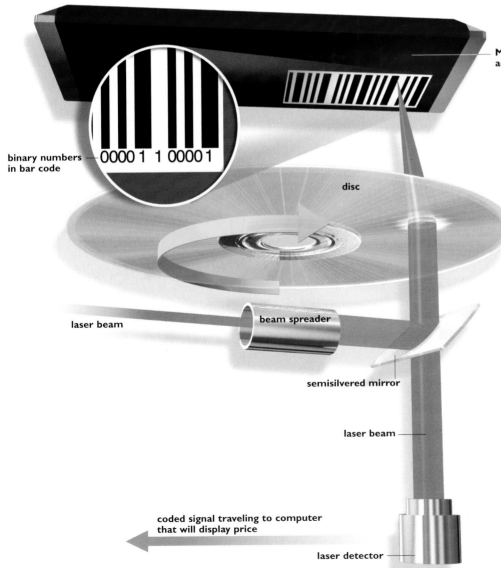

Merchandise with bar code is pulled across a checkout window.

binary numbers in bar code —— 0000 1 1 0000 1

disc

laser beam

beam spreader

semisilvered mirror

laser beam

coded signal traveling to computer that will display price

laser detector

SCANNING (left): When a clerk draws a bar-coded item over a scanner in a checkout counter, a laser beam enters a spreader and reflects off a mirror to a disc and up through the opening. The beam reads the encoded information and bounces back to the scanner's laser detector; the signal then goes to a computer that displays the price. The bars of the code represent the binary digits 0 (in white), which reflect more strongly, and 1 (in black); combinations of the units identify the item.

THE HAND CALCULATOR

SEE ALSO

**Chips and
Transistors**
• 238

Computers • 236

**How Computers
Remember**
• 240

WITH brains made of silicon chips carrying numbers coded into electronic pulses, and powered by miniscule batteries or solar cells, hand calculators have astonishing speed and accuracy. A number of interrelated electronic processes are at work, and two of the most important are the binary number system and the principles of Boolean algebra. Instead of using the based-on-ten decimal system, the binary system counts by twos (1s and 0s only).

Inside a calculator, decimal numbers and a host of functions offered on the keyboard are converted to a sequence of coded binary numbers that are stored on a memory chip. Next, Boolean algebra, named after English mathematician George Boole in the 1840s, comes into play. Essentially, it tells us that statements or values may be reduced to true or false, "on" or "off"—elements tailor-made for computing circuitry that, for all its humanlike endowments, still prefers hanging out with no-nonsense electrical pulses. Boolean principles are behind thousands of linked microscopic "switches," or logic gates, on a chip. With such Boolean names as OR, AND, and NOT, the gates (made of transistors) evaluate all the steps of calculating by switching off, say, for 0, on for 1—a judgment that is the binary equivalent of false and true—and sending these digital signals to appropriate locations. In units called half-adders that connect to form full-adders, the logic gates process signals so quickly that when the equal (=) key is pressed, the answer appears instantaneously.

Number crunchers (below): An electronic calculator and an abacus both do arithmetic without written figures. The calculator relies on a memory chip to store a mind-boggling collection of numbers and information. "Bead arithmetic" halves the time it takes to add, subtract, multiply, and divide with pencil, and in the hands of a skilled operator, it can even do complicated fractions.

BINARY					DECIMAL
8	4	2	1		
0	0	0	0	=	0
0	0	0	1	=	1
0	0	1	0	=	2
0	0	1	1	=	3
0	1	0	0	=	4
0	1	0	1	=	5
0	1	1	0	=	6
0	1	1	1	=	7
1	0	0	0	=	8
1	0	0	1	=	9
1	0	1	0	=	10

$$1 \times 8 + 0 \times 4 + 1 \times 2 + 0 \times 1$$

BINARY CODE (left): A computer or a calculator converts information to a so-called base-2 system, a binary code that relies on 1s and 0s instead of a system based on 10—the decimal system. A number's binary code equivalent uses blocks, or positions, arranged from right to left and worth 1, 2, and powers of 2 (such as 4 and 8). Building the binary 10 requires a 2-block, an 8-block, and no 4- or 1-blocks.

INSIDE A CALCULATOR (below): Beneath its keyboard panel, a calculator operates on the "yes" and "no," "on" and "off" principle that translates the binary code into matching voltage states. For example, when the binary system writes decimal 10 as 1010, it says the following: "Blocks of 2 and 8, yes; blocks of 1 and 4, no." In a calculator, switches see that 0s, which mean "no," or "off," are low voltage; and that 1s, which mean "yes," or "on," are high voltage. When a person presses decimal numbers on the keypad, a decoder produces the binary equivalents held in the storage cells.

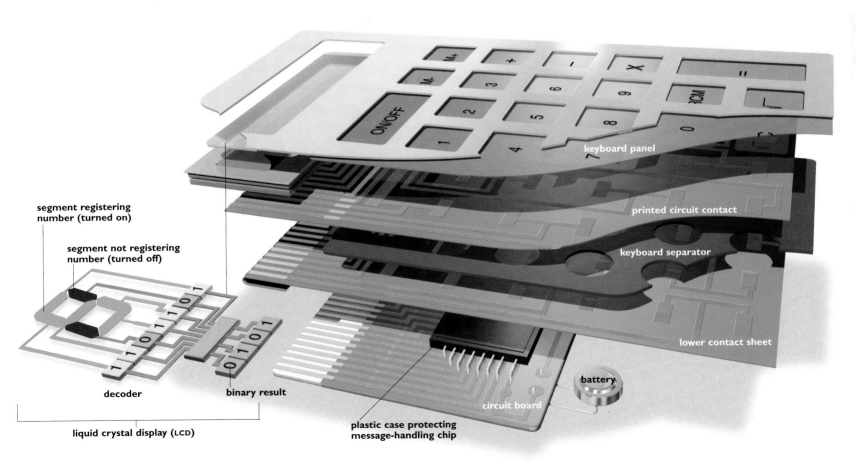

segment registering number (turned on)

segment not registering number (turned off)

decoder

binary result

liquid crystal display (LCD)

keyboard panel

printed circuit contact

keyboard separator

lower contact sheet

battery

circuit board

plastic case protecting message-handling chip

COMPUTERS: BRAINS IN A BOX

SEE ALSO

Backup Storage • 242

How Computers Remember • 240

QWERTY • 222

Weaving • 132

INSIDE A COMPUTER (below): The electronic circuitry of the central processing unit (CPU) contains arithmetic/logic and control units for calculating, transferring data, and executing instructions. Two memory units store program instructions and data for the CPU: RAM (random-access memory) holds data only while the current is on; ROM (read-only memory)—a permanent strongbox—stores digital information, such as start-up and operating programs and various computer languages and systems. An input-output unit (BIOS) processes the human-produced information and instructions and then feeds it out. Hard and floppy disks store information.

IN 1834, Charles Babbage, an English banker's son and a mathematician, conceived what he called an "analytical engine," a conglomeration of levers, gears, cogs, and wheels designed to run on steam power. Answers would be printed automatically, while the machine controlled itself internally by punched cards, an idea suggested by early mechanized looms that relied on such cards to run the pattern-weaving apparatus. Too far ahead of his time, Babbage could not drum up enough interest or money to produce his machine, but, incredibly, he had mapped out most of the key elements found in modern computers. These included an arithmetic/logic unit, memory, input and output mechanisms, and the highly imaginative notion of automatically sequencing many of the instructions by which a computer moves from one step to the next without human prodding.

Despite the computer's awesome ability to crack codes, translate languages, and solve math problems, its major advantage is speed. It is still artificially intelligent and entirely limited to what humans build into it. The billions of cells and circuits that make up the human brain's communication system far surpass all of the transistors in all of the chips in our fastest computers.

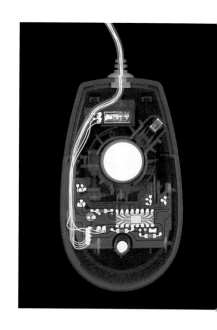

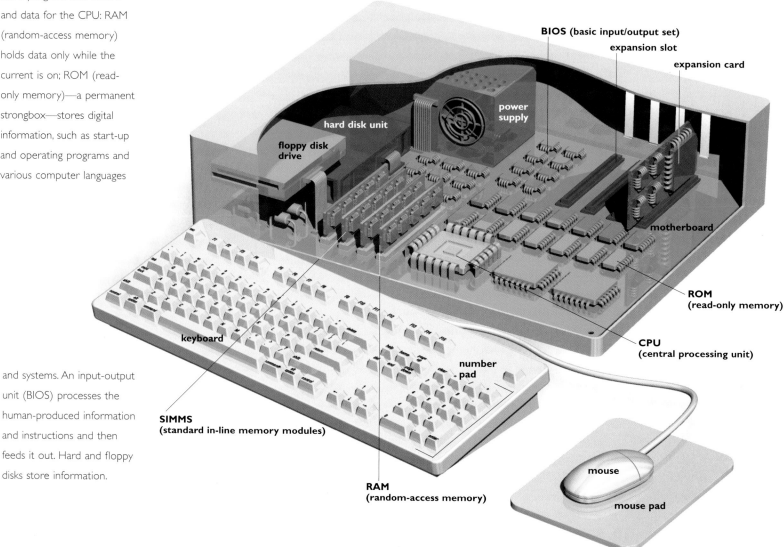

BIOS (basic input/output set)

expansion slot

expansion card

power supply

hard disk unit

floppy disk drive

motherboard

ROM (read-only memory)

keyboard

CPU (central processing unit)

number pad

SIMMS (standard in-line memory modules)

RAM (random-access memory)

mouse

mouse pad

Rolling along on a skidless ball and trailing a tail-like cable, a mouse (opposite) moves a cursor more easily than arrow keys can.

ON A ROLL (right): Moving a mouse turns the ball between three shafts. One supports the ball, and the other two, with slotted wheels, respond to movement on two right-angled axes. Photodiodes by the wheels count light pulses beamed through the slots by light-emitting diodes. The pulses move the cursor, and clicks on the mouse's finger buttons send commands to the computer.

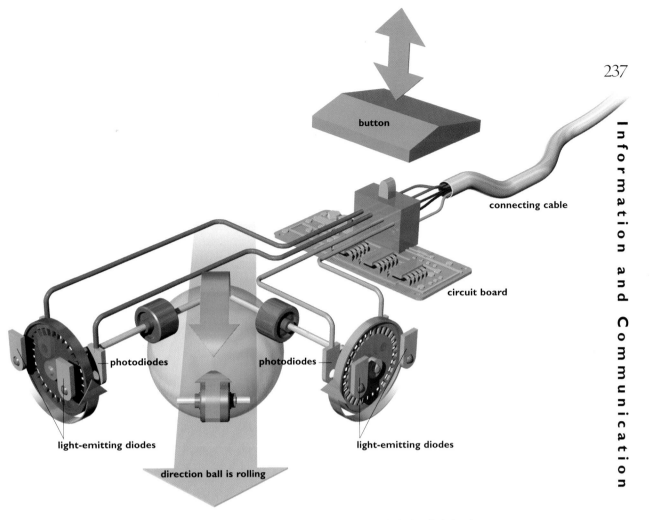

button

connecting cable

circuit board

photodiodes

photodiodes

light-emitting diodes

light-emitting diodes

direction ball is rolling

A COMPUTER KEY (left), when pressed, causes electrical contacts beneath it to close and current to flow. A signal passes to a keyboard processor that scans and sends it to the BIOS chips. After the coded signal goes to the CPU, the screen displays the pressed letter or number. The **HARD DRIVE** (below) consists of aluminum disks coated with magnetic material. Read-write heads floating over the disks convert coded electrical signals and write them in a digital code. The **HEADS** (right) are electromagnets whose north and south poles align magnetic particles on the disks to form digital 1s and 0s.

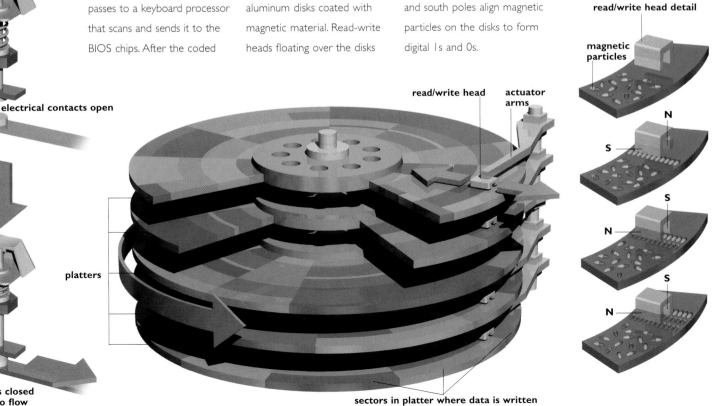

key

electrical contacts open

key depressed

platters

electrical contacts closed allowing current to flow

read/write head

actuator arms

sectors in platter where data is written

read/write head detail

magnetic particles

N

S

S

N

S

N

CHIPS AND TRANSISTORS

SEE ALSO

Backup Storage • 242

Computers • 236

How Computers Remember • 240

Radio • 148

EARLY computers relied on vacuum tubes—the airless glass "bottles" in which electric and magnetic fields control the movement of electrons—to switch electrical signals and to add, multiply, store, and compare data. Developed for the radio industry, the tubes permitted machines to calculate several thousand times faster than earlier electromechanical relays. The transistor, infinitely smaller and far more frugal with power, would put that speed to shame.

Invented in 1947 at Bell Laboratories, the transistor is an electronic device made of semiconducting material. This material, midway between a conductor and an insulator, carries current only under certain conditions, such as when a tiny amount of voltage is applied. Computer microchips, the fingernail-size flakes of silicon that control all of a computer's behavior, can each contain millions of transistors linked by fine connections to form integrated circuits. On a chip, transistors serve as electrical switches that quickly switch current off and on; such maneuvering precisely controls the flow of binary numbers, the digital data the computer uses to perform its multitude of chores. Depending on their arrangement and the purpose of the chip—memory, central processing, logic, or timing that synchronizes all the signal activity on the chip—transistors, with their ability to interpret and channel specific digital information, govern virtually everything a computer does. They control moving an icon with a mouse or deleting a letter with a keystroke; they are needed to perform complex arithmetic and to store programs and data.

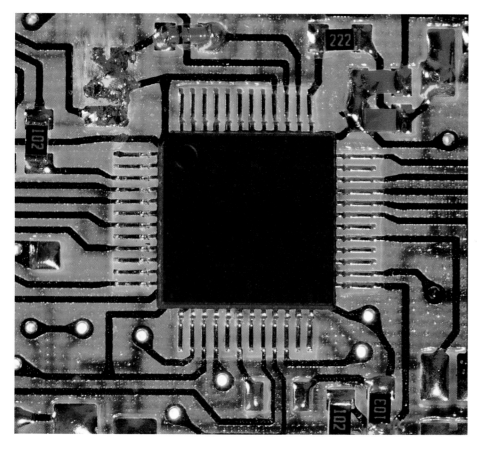

Microchip (left): Engraved with millions of circuits, wafer-thin bits of silicon contain tiny transistors that control current flow with their on-off switching capability. Microchips have different functions and a variety of arrangements on a motherboard, the printed circuitry holding components of a microcomputer system.

Microprocessor (right): Complex integrated circuitry—packed with miniature diodes, transistors, capacitors, resistors, and conductors—handles a computer's central processing. The Pentium processor shown here has millions of transistors.

238

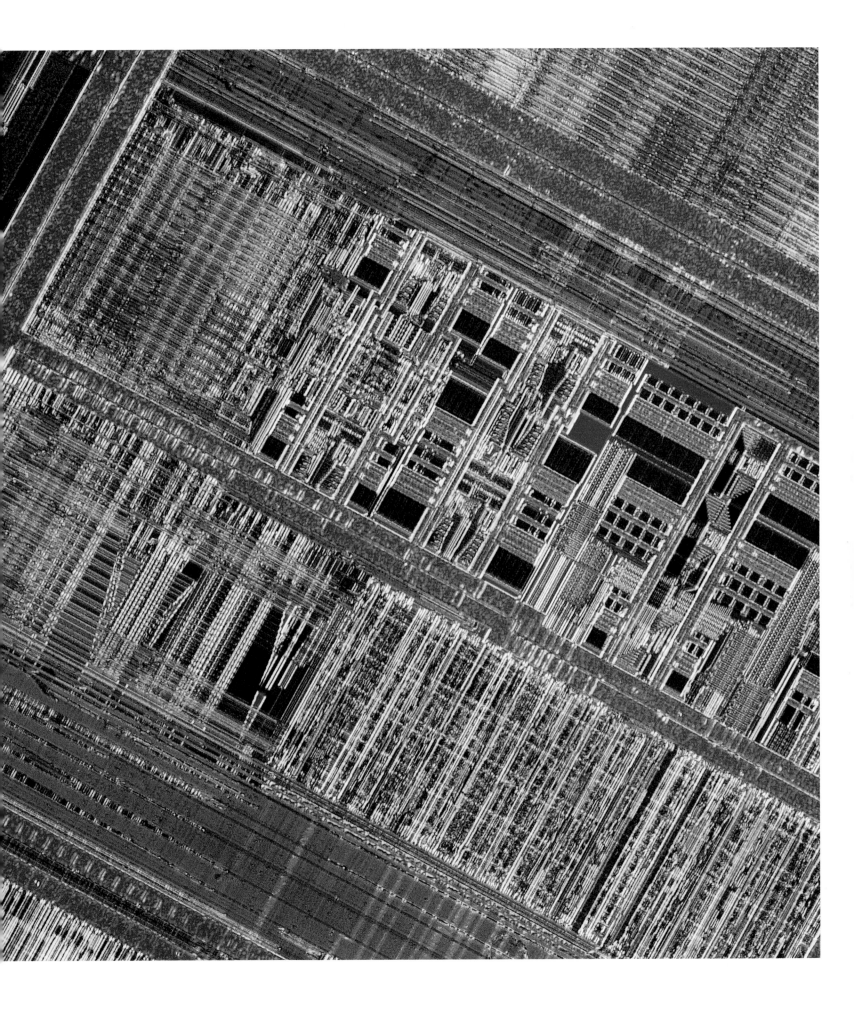

SEE ALSO

Backup Storage
• 242

Chips and
Transistors • 238

Computers • 236

Reactive robot (above): Programmed to react to sound, movement, and light, a robot demonstrates behavior-based intelligence. "IT" mimics emotions and smiles in the presence of people.

BECAUSE a computer is essentially an adding and subtracting machine, it stands to reason that it would share some of the characteristics of the electronic calculator. Both require memory to store instructions and accumulated data so that they can repeat their operations and make logical decisions, and both depend on logic gates made up of transistors to control the on and off voltages.

A computer program is a set of coded instructions that tell the machine which operations to perform and when and how to perform them. Factory- or user-installed, a program is what enables the computer to process words, play games or music, paint pictures, digitally enhance a photograph, prepare and file income taxes, and store data.

Instructions and other information must be remembered if they are to be followed, however, and computers use two kinds of memory to do that within their circuitry. Random-access memory (RAM) is where program instructions and data are stored until the central processing unit can access them. RAM is dependent on chips with memory capacity measured in accumulations of bytes—the basic units of data. (A byte is a series of 8 consecutive binary digits that can represent one from as many as 255 alphanumeric characters. Typical personal computers can have 64 megabytes of RAM, a prodigious capacity, considering that one megabyte can store a million characters.)

RAM is the computer's short-term memory. While more of it means faster computing, it holds data only when the current is switched on; hence, the reason for the cardinal rule of computing: save, save, save. ROM, for read-only memory (meaning it can be read but not altered), is the other memory bank. It is a permanent strongbox that stores digital information, such as start-up and operating programs and various computer languages and systems, even when the computer is switched off.

Gateway between physical and digital space, a metaDESK (below left) uses lenses and an arm-mounted LCD to manipulate digital information. The user interacts with two- and three-dimensional images of the MIT campus. **LOGIC GATES** (below), or transistors on chips, control binary signal flow. Gates AND, OR, and XOR switch voltage off for 0 and on for 1 (binary equivalents of false and true). Adder circuits do math: Half-adders add two binary digits; full-adders process two digits and a carry. The switching steps below add 2 and 3 in binary code to get 5.

Tangible Media Group
MIT Media Laboratory

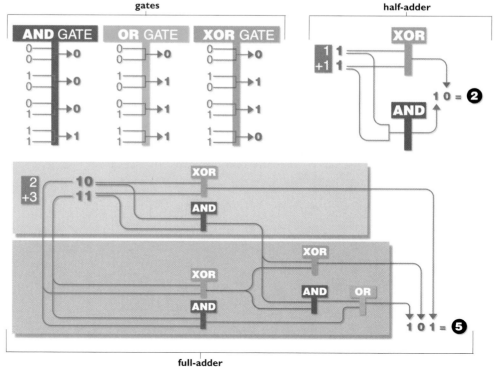

gates

half-adder

full-adder

PIONEER IN LOGIC

The ABC, an early computer that was
replicated (below) at Iowa State University
in 1998, drew on binary-coded data and
instructions arranged in a pattern of holes
on machine-readable punch cards. So-called keypunch operators made the
holes with a lever-and-pin device. Named for its builders, John V. Atanasoff,
a physicist at what was then Iowa State College, and graduate student
Clifford Berry, it was the first computer to use a binary system, parallel
processing, and other mainstays of modern computers. In the ABC,
punch cards were read (1) and translated into binary numbers (2) that
were stored on memory drums (3). Recorded on paper (4),
the binary information was interpreted by electronic
readers (5). Punch cards were first used for
large-scale data-processing by
Herman Hollerith in
the U.S. Census
of 1890.

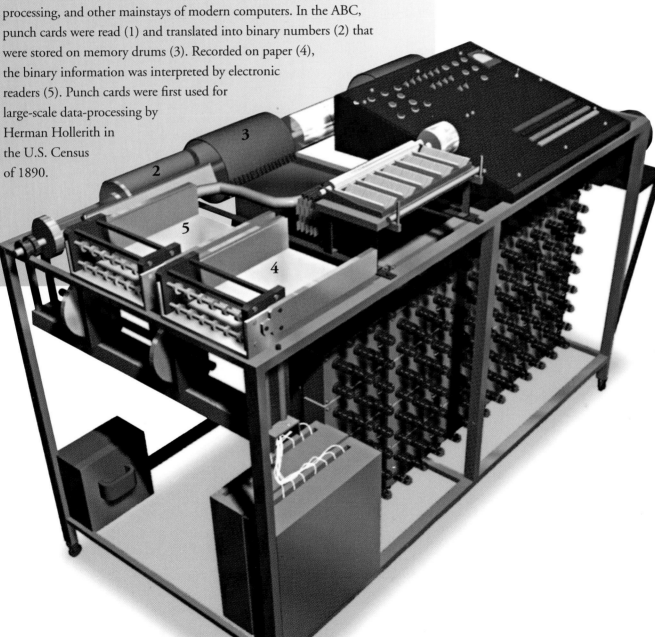

BACKUP STORAGE

SEE ALSO

Chips and Transistors • 238

Computers • 236

How Computers Remember • 240

Warhorse of data storage, the 3.5-inch floppy disk (above) remains a fixture in even the most advanced computers, even though faster readers with more capacity frequently replace it.

Superdisk (right): The removable Jaz disk system can store two billion characters in a small space. Read with magnetic heads, such superdisks work much like those in hard drives.

Shiny conveyors of sound, text, and images, CD-ROMs (far right) store a wealth of entertainment and information on digitally coded, laser-read, and big-byte-capacity disks.

BECAUSE a computer generates vast quantities of information, extra storage is essential. The depositories in general use are built-in hard disk drives that store enormous amounts of digital code, typically measured nowadays in gigabytes (a gig, as it is called, stores a billion characters, or a thousand megabytes), and removable disks.

A hard drive is actually an upgraded version of an old LP record player, only instead of just one disk, a stack of hard magnetic ones spins about in a metal enclosure. Information sent to the drive is written on the disk faces in a magnetic pattern of digital 1s and 0s by electromagnetic heads. These heads are similar to the ones in a tape recorder and can read as well as write, much as the tape recorder both records and plays back.

The familiar 3.5-inch floppy disk inserted into a computer is also magnetic, but its drive is a slower reader than a hard drive. Its heads touch the spinning disk surface, while a hard drive's heads float millionths of an inch over it; floppies also hold far less data, somewhere around two megabytes. A new generation of removable storage devices that can take anywhere from 100 megabytes to two full gigabytes has, however, nudged the floppy aside. Some devices, with storage equal to that of 70 floppy disks, are inserted directly into a slot in the computer and can store and read with magnetic heads that mimic those of a hard drive. Others, notably drives with up to two gigs of space, may require both an external drive unit that is plugged into the computer and an insertable storage disk.

Perhaps the most fascinating storage disk is the CD-ROM, a compact, laser-read disk that is just like the audio version except that it digitally stores text as well as music and images. But CD-ROMs, which can carry up to 650 megabytes of data, have the disadvantage of all ROM memory: They can only store and present information, and they cannot, for the moment at least, accept computer-written data.

SEE ALSO

Computers • 236

**Eyes and Ears
• 220**

**Sending Signals
• 150**

**Sky Telephone
• 216**

WHEN the fast riders of the Pony Express were carrying the mail, they needed eight to ten days aboard numerous horses to get a letter from Missouri to the Pacific coast. Today, even overnight mail by truck and plane seems slow, now that we have electronic mail and the Internet.

E-mail is mail sent from one computer to another via a telephone network, while the Internet over which it routinely runs is a worldwide web of computer networks that link universities, corporations, governments, commercial providers, the military, and numerous other agencies. Sending e-mail or raising an Internet site is as easy as typing a message and clicking a mouse on a send button. But the technology that gets it from here to there involves an intricate process that begins at the keyboard and includes more station stops than a Pony Express route. E-mail requires a software program that allows one to compose, send, receive, forward, and reply to text files. Like letter mail, e-mail uses addresses, most of which include the symbol "@," which separates the recipient's (and the sender's) name from his or her mail service name.

When a message is sent through the network, from computer to computer, it follows a routing program called SMTP, for Standard Mail Transfer Protocol. Messages carry an elaborate letter and number code that tells each computer station what it must do with the message so that it can reach its destination in an electronic mailbox. Messages can be delivered in seconds or minutes to computers anywhere in the world, and usually only for the price of a local telephone call.

ASTROLINK SYSTEM (below): Satellites in five positions of the Astrolink system may soon provide easier, cheaper, and additional Internet access. Placed in geosynchronous orbit, each satellite will remain above the same spot on the Earth's surface. Internet users will be able to access the system with simpler, nontracking terminals.

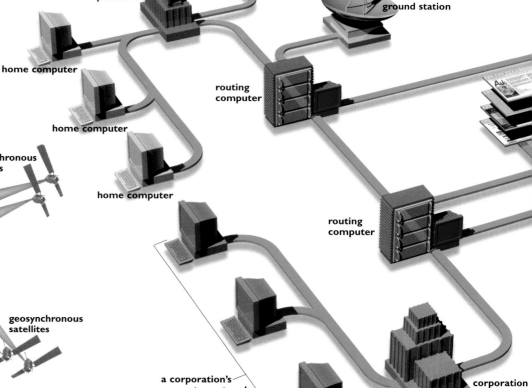

satellite

web sites

service provider

ground station

home computer

routing computer

home computer

home computer

routing computer

a corporation's computer network

corporation

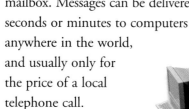

geosynchronous satellites

geosynchronous satellites

geosynchronous satellites

Earth

geosynchronous satellites

geosynchronous satellites

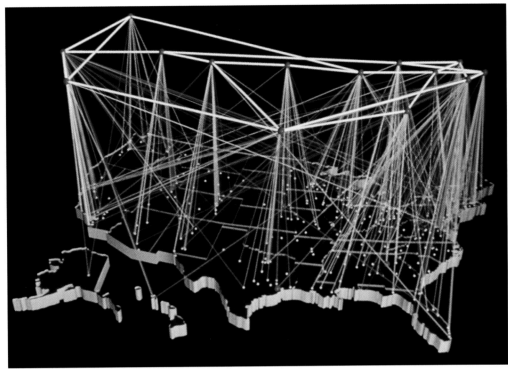

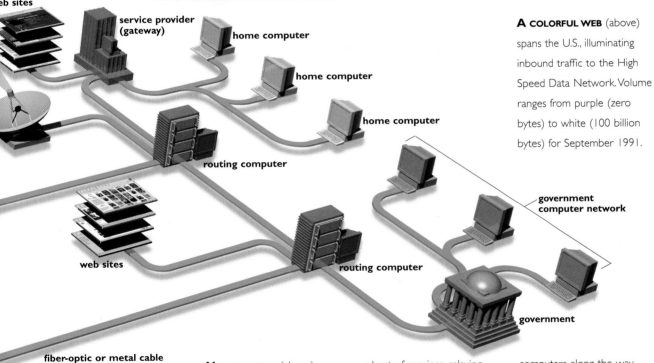

eb sites

service provider (gateway)

home computer

home computer

home computer

routing computer

web sites

routing computer

government computer network

government

fiber-optic or metal cable

A COLORFUL WEB (above) spans the U.S., illuminating inbound traffic to the High Speed Data Network. Volume ranges from purple (zero bytes) to white (100 billion bytes) for September 1991.

NETWORKING (above): A typical network connects businesses, homes, universities, and government agencies, allowing access to information. Powerful gateway computers operated by service providers connect different wide-area networks to one another, providing Internet access and a host of services, relaying digitized information and making each network's computers able to communicate with one another. Satellites and phone lines relay data and messages between computers, each one equipped with a modem or other device for transmitting and receiving data. Routing computers along the way decode instructions on the transmissions that tell them how and where to send the messages. Conceived in 1969 by the Advanced Research Projects Agency at the U.S. Defense Department, the first nationwide network linked computers at four universities.

TOOLS that let us observe realms too small or too distant to be seen by the naked eye are often baffling. Even the familiar optical devices—contact lenses, bifocals, and binoculars, for example—work in ways we may only vaguely understand. Employing lenses, mirrors, and prisms, the precision instruments we use to view microscopic or macroscopic worlds may have direct or indirect links to laws of reflection and refraction (which deal with changes in the direction of light as it passes from one medium to another). Some devices follow different laws, using lenses made from electromagnets to focus or change magnification, and beaming electrons rather than light to study objects. Others have no lenses at all and look nothing like conventional telescopes. Called radio telescopes, the dish-shaped devices receive radio waves from space, allowing us to "see" other worlds, near and far, in yet another way.

Restoring sight, astronauts of the space shuttle Columbia *visit an errant solar observatory.*

SEEING THE LIGHT

SEE ALSO

Camera • 160

Mediums and Messages • 214

Microscopes • 252

OPTICS is the science of light and vision, and optical lenses are devices that aid vision by focusing, bending, and spreading light rays emitted or reflected by an object. Unlike the lens in a human eye, which can shift its focal length through muscular contraction and relaxation, optical lenses have focal length and power built in, depending on their shape. They generally consist of two curved surfaces, or one flat and one curved; these curves may be concave (inward-curving) or convex (outward-curving). Various surfaces and thicknesses determine a lens's focal power and function, while combinations of different lenses—the compound lenses cemented together in a simple light microscope, for example—prevent the blurring, distortion, and other anomalies that can occur with single, thin lenses.

Eye specialists have many techniques available to them for measuring the eye's refractive power and for determining a person's need for corrective lenses. They also have several ways of dealing with vision problems. To compensate for nearsightedness, for example, lenses are ground in concave shapes; for farsightedness, lenses are convex. Cylindrical lenses are used for astigmatism, a condition in which light does not focus properly on the retina because of a defect in the curvature of the natural lens. Prisms, which bend, spread, and reflect light, are used for other defects.

Combinations may be required. In bifocals, the upper and lower parts of a lens are ground differently to correct for both close and distant vision. Trifocals are ground with a center lens for intermediate distance. So-called occupational lenses, often worn by computer-users, correct for near vision at eye level and for other distances above and below.

Eye contact (above): Made of shatterproof, soft plastic, a contact lens fits over the cornea, the outer, transparent part of the eye.

Through a looking glass (opposite): A piece of high-tech equipment receives a visual inspection in a modern manufacturer's quality-control lab. Complex creations in convex and concave forms, lenses compensate for inadequacies in human eyes.

Binoculars (above) give a wildlife-watcher a steadier view than a handheld spotting scope, which trembles even at low magnification.

BINOCULAR VISION (right): To see, human eyes form two images of an object—one on each retina. Binoculars accommodate that by letting light enter through two large objective lenses. Reflecting prisms inside each tube fold the light's path, providing higher magnification than that afforded by field glasses, which have no prisms. The prisms also allow for a more compact binocular design; field glasses need much longer and heavier tubes for high magnification.

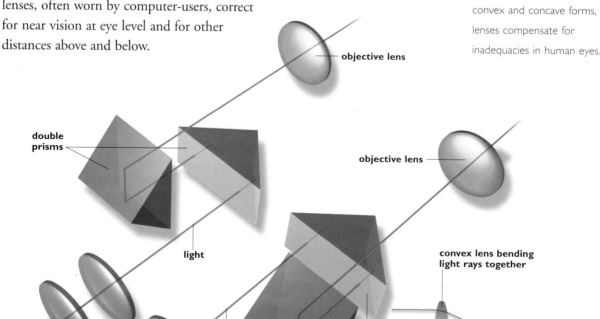

objective lens

double prisms

objective lens

light

light

convex lens bending light rays together

ocular lens system

ocular lens system

double prisms

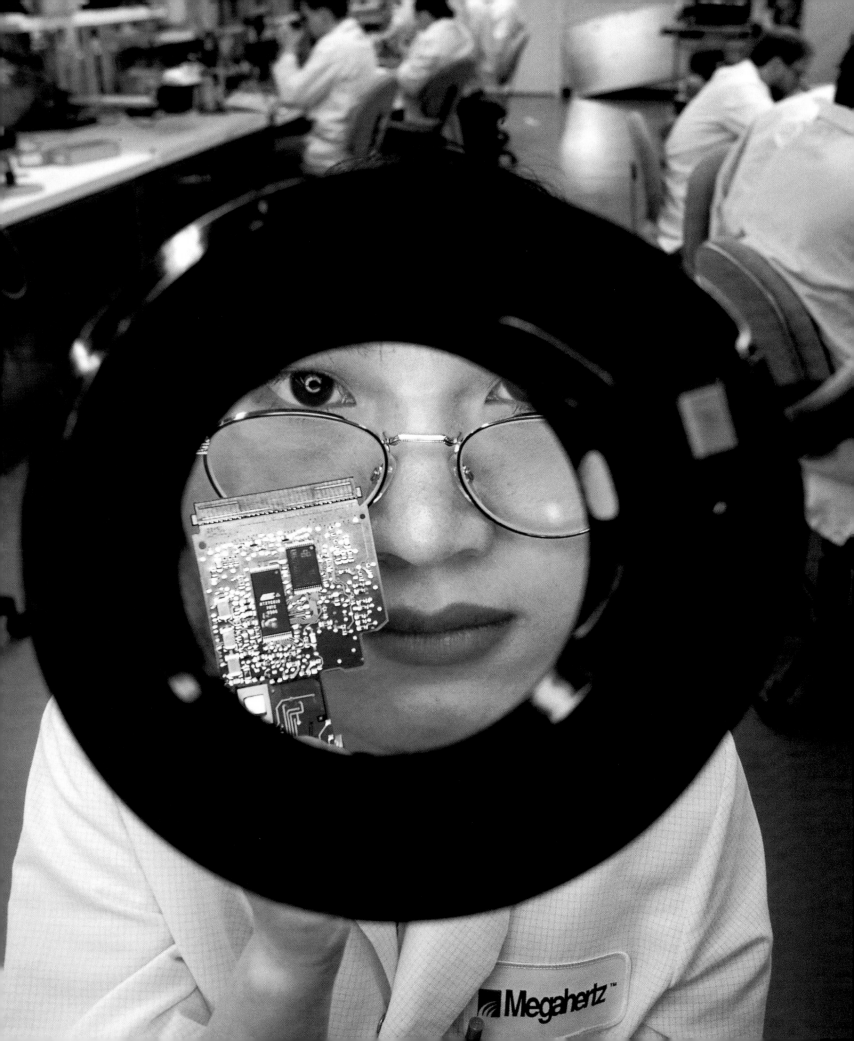

NIGHT VISION

SEE ALSO

**Mediums and
Messages • 214**

**Seeing the
Light • 248**

**ELECTRO-OPTICAL NIGHT
VISION** (below) works by
intensifying and increasing
available light. As faint light
enters a photocathode, its
photons—a stream of energy
packets—accelerate through a
vacuum. Next, a microchannel
plate multiplies the photons
thousands of times. The device
converts them to an electronic
image and focuses it on a
phosphor screen for viewing.
Amplified green light then
enhances the image's visibility.

WHAT does it take to see in the dark? Humans and other mammals need the
rods in the retina, millions of tiny cylindrical elements that contain rhodopsin, a
purple pigment that can detect dim light. We also have millions of retinal cones,
light-sensitive cells that enable us to read fine print. Cats and owls, however,
have only rods, a choice of nature that makes them nocturnal creatures capable
of seeing far better than humans in low light.

For a human eye to see in a dark movie theater, its rods have to rejuvenate
rhodopsin, which is so chemically altered by bright outside light that the rods
temporarily lose their sensitivity to light. While rhodopsin's response allows
the eye to withstand the sunlight, it doesn't help much when you enter a dark
screening room. Gradually, however, more rhodopsin is produced, increasing
sensitivity to the theater's low light.
Vitamin A, a lack of which can
result in night blindness, is essential
to the production of rhodopsin.

Medical science may someday
create a pill that allows us to see
better in the dark, but until that
time, technology must fill the bill.
Night-vision binoculars and scopes
are electro-optical instruments that
are incredibly sensitive to a broad
range of light, from visible through
infrared. Light that enters a lens in
a night-vision scope reflects off an
image intensifier to a photocathode
and is converted to an electronic
image. Amplified on a viewing
screen, the image reveals much
more than a night scene observed
through a conventional scope.

Working in the dark, and
doing it efficiently, a member
of a French special police
force (below) peers through
the blackness with night-vision
goggles. Such devices require
some ambient light, perhaps
from the stars or from a
focused-beam infrared source.
Like cameras, the goggles have
various image magnifications.

amplified green light

what the eye sees   phosphor screen

photons (light energy) converted
to electrical energy or electrons

microchannel plate   photocathode

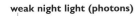
weak night light (photons)

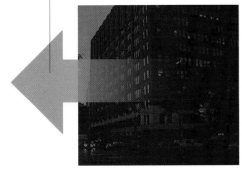

Captured: A perpetrator can run, but he can't hide in a New York City subway station (above). Rendered an eerie, all-seeing green when viewed by night-vision goggles, such images get brightness from photocathodes that are highly sensitive to infrared light. Because the naked eye cannot see infrared light, people who illuminate scenes with such light may remain unseen by the subjects they observe. The strength of both the magnifying lens and the image intensifier determines the distance from which one person can clearly recognize another. Some civilian models can view objects 400 yards away. Police, military personnel, and many civilians rely on night-vision technology, including boaters who use the devices to locate buoys and moorings and to detect other vessels.

LIGHT AND ELECTRON MICROSCOPES

SEE ALSO

Mediums and Messages • 214

Seeing the Light • 248

Television • 156

X-Rays, CT, and MRI • 196

An electron microscopist (right) gets finer details than through the more familiar eyepiece and tube of a light microscope. Equipped with a monitor and using electron beams, not light, the electron microscope draws more on television technology than on the lens science behind conventional microscopes. Indeed, electromagnets, rather than glass, function as the lens. Electrons pass through thinly sliced specimens coated with gold or water vapor to improve the image.

"WHERE the telescope ends," wrote Victor Hugo in the novel *Les Misérables,* "the microscope begins. Which of the two has the grander view?"

As nearly everyone knows, microscopes are instruments that give us magnified images of very small objects. But when Hugo penned those words in the 19th century, he was speaking of the simple light microscope with its then astonishing, and now very limited, view. Today, the telescope has not ended at all, and the microscope has long said goodbye to its beginning. Both are still evolving and producing startling images.

Microscopes come in many sizes and shapes. Perhaps the most familiar is the ordinary compound microscope, which has a pair of convex lenses of short focal length on the lower (objective) end of a tube and another pair at the eyepiece. When a specimen is placed at the objective end and illuminated with light reflected from a mirror, the magnified image is magnified again at the eyepiece end. Focusing is done by moving the objective lenses nearer to, or farther from, the specimen.

While the light microscope can identify the form and structure of extremely tiny organisms, the electron microscope reveals far more detailed information about their surface and inner workings. Under an electron microscope, a simple bacterium bares its very soul and becomes an intricate cutaway of the complexity of life. With a magnification power of up to hundreds of thousands of times, the microscope can see objects that are among the most invisible of the invisible: atoms. Armed with an electron gun, this nonlight microscope focuses a beam of electrons through a vacuum and over the surface of a specimen. A signal is generated, projected onto a fluorescent screen, and then photographed. Instead of optical lenses, the microscope uses magnetic lenses to focus and change magnification. These produce a field that acts on the electron beam in the same way a glass lens works on light rays. The exceptionally high resolution is due to the shorter wavelength associated with electron waves.

A SCANNING ELECTRON MICROSCOPE (opposite) relies on a beam of electrons produced by a heated filament and accelerated under high voltage. Fired by an electron gun, as in a TV set, the beam passes down the microscope's column through a vacuum. Bent by electromagnets, much as glass lenses bend light, the beam focuses on a specimen cut thin to allow the electrons (continued at top of p. 253)

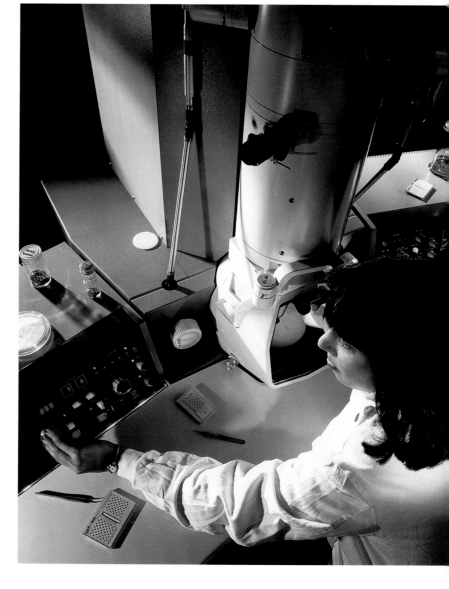

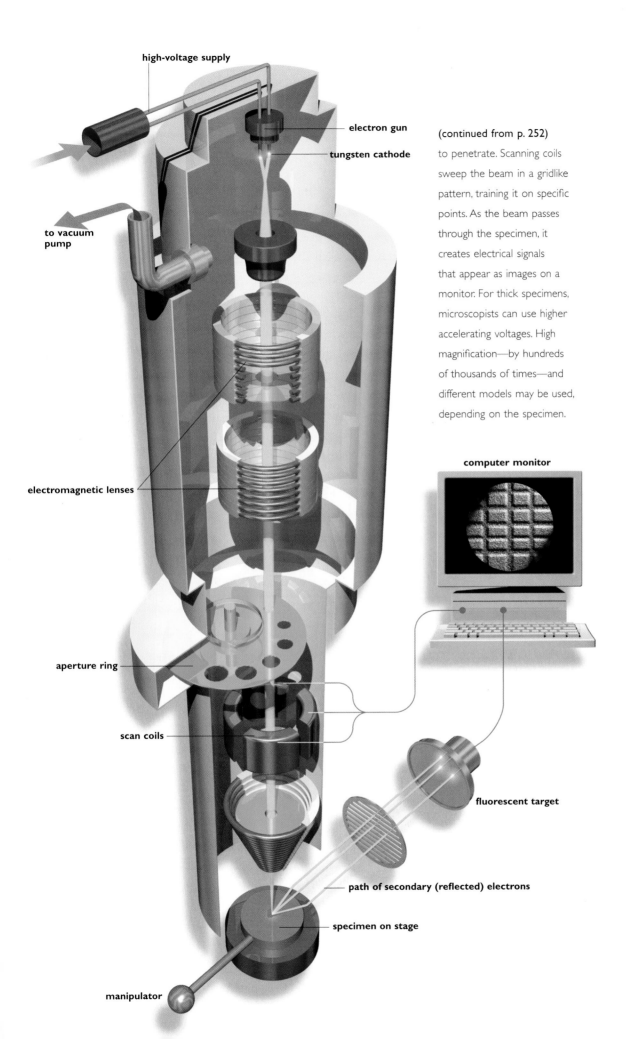

high-voltage supply

to vacuum
pump

electron gun

tungsten cathode

electromagnetic lenses

aperture ring

scan coils

manipulator

computer monitor

(continued from p. 252)
to penetrate. Scanning coils
sweep the beam in a gridlike
pattern, training it on specific
points. As the beam passes
through the specimen, it
creates electrical signals
that appear as images on a
monitor. For thick specimens,
microscopists can use higher
accelerating voltages. High
magnification—by hundreds
of thousands of times—and
different models may be used,
depending on the specimen.

fluorescent target

path of secondary (reflected) electrons

specimen on stage

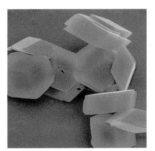

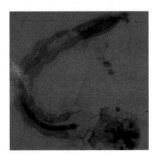

Insulin crystals (upper) and
mosquito larvae (lower),
magnified by magnets, reveal
their complex structure in
exquisite detail when viewed
by electron microscopes.

SEE ALSO

Mediums and Messages • 214

Seeing the Light • 248

Space Telescopes • 256

Polishing the mirror of a giant telescope (above) provides more than superficial shine. It's necessary to remove pits and surface defects that can distort images.

REFLECTING TELESCOPES (below) use curved mirrors to magnify distant objects. In the so-called Cassegrain reflector, a 17th-century creation, light passes through a hole in the main mirror and focuses the image on a secondary mirror. To view the image, one peers through the eyepiece. William Herschel's 1773 reflector used a tilted main mirror with an eyepiece looking into it. Isaac Newton's model, only six inches long, had two mirrors; coudé versions use three mirrors to focus images to the side.

GALILEO did not invent the telescope, but he was arguably the first person to use one for making serious astronomical observations. With his 17th-century device, he found moons around Jupiter, mountains on Earth's moon, and spots on the sun. Galileo's telescope was a simple arrangement of lenses in a tube that, for all they accomplished, bent light at odd angles, thus distorting color and producing images of poor quality. Later, Isaac Newton rectified that, using mirrors to reflect light rather than bend it. Modern astronomers often view an image on a screen or as a photograph, without actually looking through an eyepiece.

Telescope mirrors are massive constructions of polished glass having diameters of up to 32.2 feet. Spun-cast in rotating furnaces from tons of raw glass, they are linked to computers and photodetectors that, in turn, link multiple telescopes. The mirrors have enormous light-gathering power and can view wide areas of the sky for large-scale surveys of the faintest objects in deep space. Indeed, instruments now under development will allow simultaneous observation of the spectra from as many as 300 galaxies. Just as important, they are distortion-free, cast as they are from vastly improved glass mixtures and finely polished by beams of electrically charged atoms. Computers also eliminate atmospheric distortion by analyzing light and correcting it if necessary. New optical detectors promise even more: Their sensitivity allows them to clock the arrival of a single light particle and to measure its energy with exceptional precision—all through the infrared, optical, and ultraviolet segments of the spectrum.

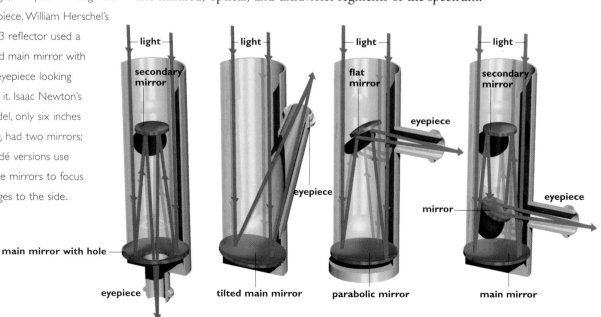

Cassegrain reflector

Herschel's reflector

Newtonian reflector

Coudé telescope

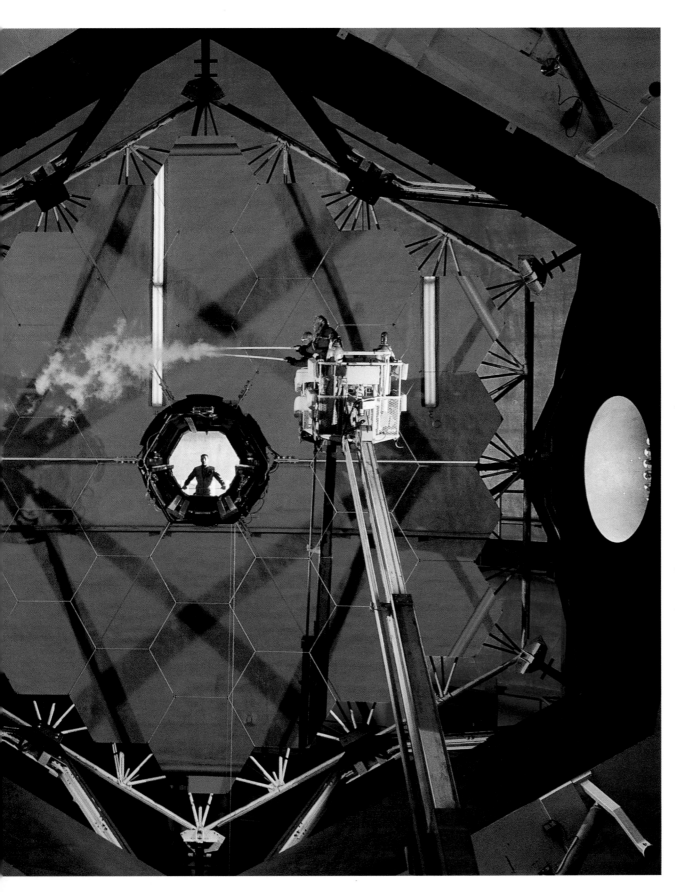

Eye on the universe (left): The massive Keck Telescope, mounted 13,600 feet high on Hawaii's Mauna Kea, receives a cleaning with carbon dioxide snow. The 32.2-foot-diameter mirror contains 36 hexagonal segments, each of which can be aimed by computer. In this view, the mirror reflects an inner wall of its dome, hence the orange appearance. The opening at the center holds instruments for observation.

SPACE TELESCOPES

SEE ALSO

Eyes and Ears
• 220

From the
Ground • 254

Mediums and
Messages • 214

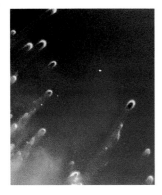

Tadpole-shaped objects

(above) in the Helix nebula,

each twice as wide as our

solar system, appear in an

image made by the Hubble

Telescope. Over the years,

the familiar rings of **Saturn**

(below) showed up in

numerous photographs and

caricatures, but sharp images

of auroras in the planet's polar

regions and the atmosphere's

fluorescent-like glow had to

await the discerning ultraviolet

gaze of Hubble.

WHILE ground-based observatories keep expanding their ability to see into deep space, there is, as the saying goes, nothing like being there. Spurred off-planet by atmospheric interference, orbiting observatories can see into radiation wavelengths difficult to image by Earth-based instruments. One space telescope examines high-energy processes in the nuclei of galaxies and in the vicinity of black holes, and another such telescope detects the most penetrating of radioactive emissions, the gamma rays spawned by violent supernova explosions.

One of the most powerful space telescopes bears the name of Edwin Powell Hubble, the U.S. astronomer who more than 70 years ago discovered that certain nebulae are really galaxies outside our own Milky Way. Hubble found that the entire universe is expanding, rushing outward from what may have been the big bang of creation. Launched aboard the space shuttle *Discovery* in 1990, the Hubble Space Telescope is a 43-foot-long, high-resolution, mirror-carrying instrument that can see objects a hundred times dimmer than any detectable by telescopes on Earth. In orbit 370 miles up, Hubble gathers light from a large, concave primary mirror and reflects it off another mirror into an array of sensors. Sensitive instruments detect x-rays, infrared light, and ultraviolet light, revealing the makeup of far-off celestial structures and systems. Hubble has found evidence of a black hole and shown us Saturn's auroras. By focusing on a seemingly empty bit of sky, Hubble has even discovered what may be a "construction site for galaxies," a vast birthplace of stars, 12 billion years in the past, that had been invisible to astronomers.

Yellow stars, millions of years

old, stand out from massive,

relatively young white stars in

this view (right) of a star cluster

in the Large Magellanic Cloud.

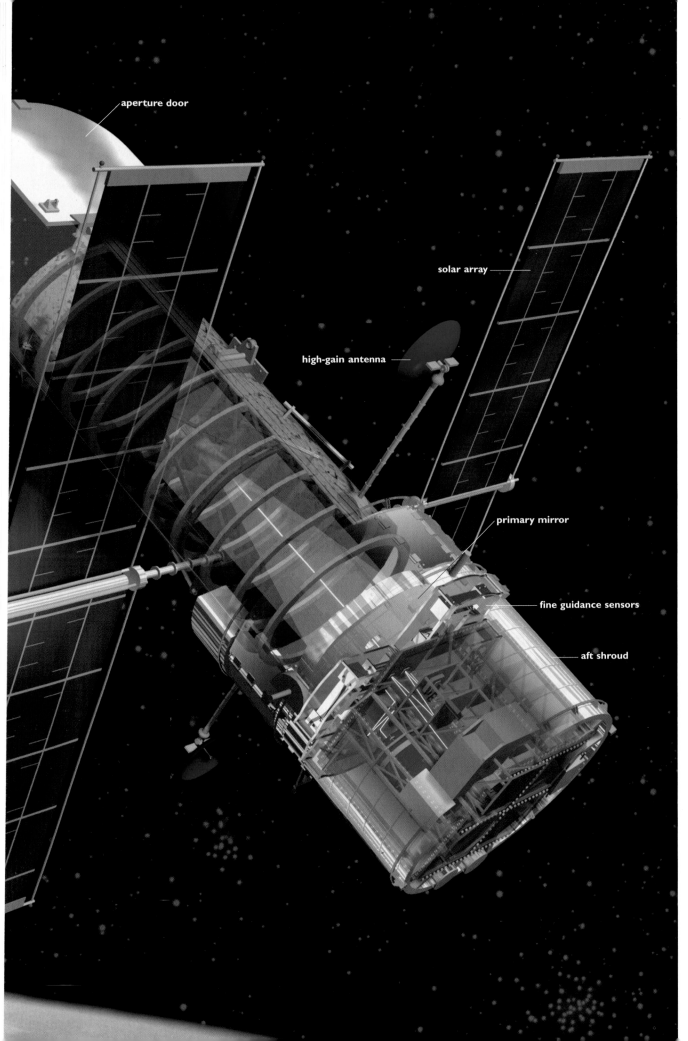

aperture door

solar array

high-gain antenna

primary mirror

fine guidance sensors

aft shroud

WITH ITS HUGE MIRRORS, cameras, and spectrographs seeking, and sometimes finding, distant glimmers from the big bang, the Hubble Space Telescope converts faint starlight into limelight. It orbits above the atmosphere, providing images free of the distortion seen in views from Earth-based telescopes not equipped with computer-controlled adaptive optics. The telescope's imaging spectrographs can read light emitted by gases, depicting it as red for nitrogen, green for hydrogen, and blue for oxygen —for all the world to see.

270

Osler, Sir William 208
Otis, Elisha 74
Ovens 15, **16;** convection 16; electric **16, 21;** gas 16, **17, 25;** microwave 16, **17;** pilot lights 16, 17, 25; toaster 16

Pagers 218, **218**
Paper: giant roll **180;** made from wood pulp 180, **180;** recycling 192; reproduction *see* Photocopying machines; Printers
Penfield, Wilder 261
Pentium processor 238, **238–239**
Petrochemicals 188; plant **188–189**
Petroleum 188; Alaska pipeline 188; being pumped **188;** distillation 188, **189;** drilling rigs 188; embargo (1973) 54; *see also* Plastics
Pezzetta, Roberto 19
Pharmaceutical companies: control room **208;** equipment **209;** research and development 208
Photocopying machines 226, **227;** Xerox 226, **226;** *see also* Printers
Photography *see* Cameras
Pianos 8, 146; concert grand **147;** keys 6, 146, **146–147;** mechanism **146,** 147
Pigs: genetically altered 124, **124–125**
Pipes 15, 22–23, 188; drainpipes 20, 22–23, **23;** heating 24, 26; water outlet tubes **56–57**
Plastics 13, 139–140, **190–191;** cards 230, 231, **231;** disposal problems 190; properties 190; recycling 192; thermoplastic and thermosetting 168, 190
Plexiglas 190
Plumbing 22, **22–23,** 34
Polar fleece **139**
Policemen: bullet-resistant vests 140; motorcycles 88; night-vision goggles **250,** 251
Polyester 138, 140, 170; recycled **139,** 140
Polyethylene 138, **190–191**
Polymers 131, 138, 140, 190–192, 202
Polyurethane 136, 138, **140,** 168
Pope John Paul II 259

Powell, James 96
Power lines, overhead 30, **30, 52–53**
Power plants 54, **55, 62–63;** coal-burning 54–55; control room **55;** giant turbines **54;** substations 55
Printers 228, **228–229;** computer **228–229;** dot matrix **229;** ink-jet 228, **229;** laser 228, **228;** printing presses 224, **224–225;** toner 228
Prosthetics 202; hip joint **203**
Pulleys 7, **7–9**
Punch cards 134, 236, 241, **241**
Pyrotechnics *see* Fireworks

Qiana 138

Racing, automobile: video game 166, **167**
Radar 114, **114,** 116, 214
Radiators: convector units 24; steam **24**
Radio waves 148, **148,** 150, **151,** 182, 197, 211, 212, 214, **215, 216,** 216–217, 247
Radios 12, **148–149,** 150, 156; dial **148;** frequencies 148, 150–151, **151;** micro-size 143; transmission towers 150, **150–151;** tuning circuit 149, **149**
Railroads: percentage of freight carried in U.S. 92; railway tracks **98;** signalmen 98, **99;** yard **92**
RAM (random-access memory) 236, **236,** 240
Rayon 138, 180
Records: listening booths **152;** long-playing 152, 242; styluses 152
Recycling 139, **139,** 140, 192, **192–193;** plant **193**
Refrigerants 28, 29; *see also* Coolants
Refrigerators 10, 15, **18,** 19, 29; compressors 18, **19;** Frigidaire **18;** frost-free 18; Oz model **19**
Reliance Building, Chicago, Ill. **70**
Reynolds, Sir Joshua 179
Rhodopsin 250
Robots **48, 84–85, 117,** 195, 204, **240**

Rocket ships **117**
Roentgen, Wilhelm Conrad 196
Roller coasters 143, 174, **174–175;** relative safety 174
ROM (read-only memory) 236, **236,** 240
Romans, ancient: aqueducts 66; chain mail 140; glass-making 182; heating systems 24; plumbing 22
Rominger, Richard 120
Rope 7, 138, 140

Sailboats **100–101;** hand crank **100;** lift, drag, and thrust 100; rigging 100, **100–101;** square-riggers **100;** tacking **100**
Saji, Yoshiro 104
Sakharov, Andrei 60
Sandburg, Carl 70
Satellites 30, 216; communication 220; geosynchronous **244;** Global Navigation 221, **221;** Global Positioning System 119, 120, 220, **221;** Iridium 220, **220**
Saturn (planet) **256**
Saws 46, **46–47**
Saxophone **144**
Scanners 11, 13, 218, 232, **232, 233**
Scanning electron microscopes 252, **253;** computer monitor **253**
Sea Cloud (sailboat) **100**
Security systems: burglar alarms 42–43, **43;** fire alarms 72; pin-tumbler cylinder locks 42, **42;** smoke alarms 42, **42**
Sewing machines 134, **134–135;** bobbins 134, **135;** feed dogs 134, **135;** lockstitch 134, **135;** needles 134, **135;** Singer 134, **135**
Shakespeare, William 132
Shavers, electric 20, **20**
Shaw, George Bernard 261
Sheep: cloned **124**
Shetland wool 139
Ships **112;** construction 102, **103;** digital model **165;** factory ships 126; hulls 102, **103;** maglev 104, **105;** simulator 102, **102;** stabilizers 103, **103;** water displacement 102–103, **103;** *see also* Nuclear submarines

Sholes, Christopher Latham 222
Sikorsky, Igor 112
Silicon chips 166, 230, 234; *see also* Microchips
Silk **130–131,** 131, **139;** artificial 138, 180; spun by moth larvae 139
Singer, Isaac M. 134
Sinks, bathroom 22, **22**
Skates, roller: one-horsepower engine **88**
Skating rinks: resurfacing 172, **172–173**
Skull, human: MRI scan **194–195;** x-ray **196**
Skyscrapers 66, 68, 70, **70–71,** 72, 74; construction worker **71;** curtain wall facades **70;** glass-sheathed 70, **71**
Soda bottles: recycling 139–140, **192**
Soil: monitoring of 120, **121;** zeolite 128
Solar energy 26, 53; heating 26, **26–27;** solar panels **26**
Sonar (sound navigation and ranging) 107; diagram **107**
Sound, speed of 116
Sound waves **145,** 148, 150, 156, 211, 212; frequencies 144
Space shuttles 220, 256; astronauts **246–247**
Spacecraft **117, 246–247**
Spinning 131–132, **132, 138;** *see also* Weaving
Sports **168–172**
Sputnik I (satellite) 220
St. Lawrence River, Canada-U.S.: ice patrol **112**
Standard Mail Transfer Protocol (SMTP) 244
Stanford, Leland 164
Stanley, William 54
Stars **256;** black hole 256; nuclear fusion 60
Steel 13, **64–65, 70;** coils **186;** furnaces **178–179,** 186, **186–187;** molten **178–179, 186;** recycling 192; stainless 186
Stores, retail: bar codes 232
Stoves *see* Ovens
Street cars: pantograph **94**
Strongboxes 230, **230,** 240
Stunt pilot **109**
Styrofoam 190
Submarines *see* Nuclear submarines
Superconductors **97,** 104

Supermarkets: bar codes 232, **232;** cash registers 232
Surgery: hip-joint replacement **203;** keyhole **198;** laser 206, **206–207;** minimally invasive **198,** 199, **199;** robots 204; virtual reality operation **204**
Synthetic Bovine Growth Hormone (BGH) *see* Bovine Growth Hormone
Synthetics *see* Fibers, synthetic; Plastics

Tamm, Igor 60
Tape recorders 152, 158, 242
Technology: advantages and disadvantages 259–261; protests against 258
Teflon 190
Telephones 6, 8, 212, **216–217,** 218, 220, 244, 245; automated switching equipment 212, **212;** booth **210;** cellular 216, **216–217,** x-rayed **217;** cordless 212, 216, **216, 217;** digital 217; fiber-optic equipment 212, **214;** micromechanical resonator **217;** switchboard operator **213;** Touch-Tone dialing 217; transmission of calls 213, **213,** 216, **216;** videophone 219, 260
Telescopes, space 247, 252, 256, **256–257;** giant 254, **254–255;** reflectors **254;** Very Large Array (VLA) 221, **221**
Television 156, **157,** 220, 261; digital **156;** high-definition (HDTV) 156; laser components **156;** monitors 158; picture tube 156, **157;** Trinitron system 156, **157**
Tennis: balls 168, **169;** racquets 168, **169,** 170
Textiles **130–131, 137,** 138, **139;** animal and plant fibers 131; bolts of cloth **132, 137;** cellulose-based 180; fireproofing 140, **140;** high-tech **140;** microPCMs 140; polyurethane-coated **140;** printing of patterns 136, **136–137;** reinforced 140; thermal **139,** 140; yarn **132–133;** *see also* Fibers, synthetic
Thermostats 16, 22, 72

Thompson, Benjamin 24
Thompson, LaMarcus 174
Titanic (film) 143; digital model **165**
Titanium 168, **168,** 176
Toasters, electric **16**
Toilets, flush **22;** intake system 22
Tokamak (nuclear fusion device) **60–61;** superconducting coils 60, **60**
Tomlin, Lili **213**
Torque 7, 44, 110
Trains **13, 80–81, 98, 184;** diesel 12, **13,** 90, 92, **92–93,** 98; electric 90, 94, 98; engine driver 90, **91;** freight cars **92;** gauges 90, **91;** maglev 96, **96–97,** 98; Metroliner 96; monorail 94; signals and traffic control 94, 98, **98–99;** steam locomotives 7, 90, **90–91,** 92; subway **94,** 96
Transformers 15, 54–56, **57**
Transistors **155, 238, 239,** 240, 260
Transmission towers 150, **150–151**
Transplants, surgical 201, 202, **202**
Trees: felled for lumber **180–181;** planting of seedlings 180
Tritium 60
Trumpets: piston action **145**
Tunnel-boring machines (TBMs) 76, **77**
Tunnels *see* Chunnel
Turbines 8, 56, 106, 110, **111;** giant **54;** steam 7; wind 62, **62**
Turkey: solar water heaters on roof **26**
Twain, Mark 261
Typewriters 8; electric 222, **223;** manual 222, **222, 223,** 258; QWERTY configuration 222; Selectric **223;** touch typing **223;** typing heads 222, 223

Ulsan Hyundai shipyard, South Korea: shipbuilding 102, **103**
Ultraviolet rays 214, **215**
Uniform Product Code (UPC) 232
United States: annual number of prescriptions 208; communications network **245**

Uranium 58–59, **59**
U.S. Census (1890) 241
U.S. Golf Association: Research and Test Center **171**
USS *Billfish:* control room **107**
USS *Nashville:* helicopter landing on **112**

Vacuum cleaners 10, **11;** canister 36; central 36; stroboscopic photo **36–37;** suction power 36; upright 36, **36**
VCRs *see* Videocassette recorders
Vegetables 261; biotechnology 124; cucumbers 128, **129;** hydroponics 128, **129;** tomatoes **120,** 124, **129**
Velcro 11, 140, **141**
Very Large Array (VLA) Radio Telescopes 221, **221**
Video cameras **158**
Video games 143; 3-D **1,** 166, **166;** benefits to children 166; electronic displays 166, **167;** game concepts 166; interactive **1,** 166, **166**
Videocassette recorders 158, **159;** magnetic tape **159**
Virtual reality: surgical operation **204;** video games **166, 167**
Volts and voltage 30–31, 44, 54–55, 94, 238, 240, 253

Washbasins 22, **22**
Washing machines 15, **34,** 35; agitators 34, 35, **35;** cleaning

agents 10, 34–35, **35;** front-loaders 34; rotating drums 34; top-loaders **35**
Washington, D.C.: subway train **94**
Waterpower *see* Hydroelectric power plants
Way, Stewart 104
Weather: forecasting 220
Weaving 131, **133,** 136, **137;** warp and weft threads 132, **133,** 136; *see also* Spinning
Wheels **8,** 9; automobiles 8, 82, **82,** 83, 85; clocks 38, **39;** Ferris **8;** pulleys **7–9**
Whitbread Race 100, **101**
Whittier, John Greenleaf 122
Wind turbines 62, **62**
Wood 180
Wood pulp **180;** cellulose 138; paper production 180, **180**
Woodwinds 144–145, **145**
Wool 131, **137, 139;** chemically treated **140;** synthetic 190
Wright brothers 110

X-rays 196, **196,** 214, **215, 217**

Yeager, Chuck 116

Zamboni (ice resurfacing machine) 11, 172, **172–173**
Zippers 12, 134, 138; teeth **134**
Zworykin, Vladimir Kosma 156

Library of Congress Cataloging-in-Publication Data

Langone, John
    National Geographic's how things work: everyday technology explained / John Langone.
       p. cm.
    ISBN 0-7922-7150-5 (alk. paper). -- ISBN 0-7922-7151-3 (alk. paper)
    I. Technology--Popular works. 2. Inventions--Popular works.
    I. National Geographic Society (U.S.) II. Title. III. Title: How things work.
    T47.L29 1999
    600--DC21   99-11776

Composition for this book done by the National Geographic Society Book Division. Printed and bound by R. R. Donnelley & Sons, Willard, Ohio. Color separations by North American Color, Portage, Michigan. Dust jacket printed by the Miken Co., Cheektowaga, New York.

National Geographic's How Things Work: Everyday Technology Explained

By John Langone
Art by Slim Films, Inc.

Published by the National Geographic Society

John M. Fahey, Jr.	*President and Chief Executive Officer*
Gilbert M. Grosvenor	*Chairman of the Board*
Nina D. Hoffman	*Senior Vice President*

Prepared by the Book Division

William R. Gray	*Vice President and Director*
Charles Kogod	*Assistant Director*
Barbara A. Payne	*Editorial Director and Managing Editor*
David Griffin	*Design Director*

Staff for this book

Martha C. Christian	*Editor*
Carolinda E. Averitt	*Text Editor*
Greta Arnold	*Illustrations Editor*
Cinda Rose	*Art Director*
Elisabeth B. Booz, Kimberly A. Kostyal	*Senior Researchers*
Joyce Marshall Caldwell, Alexander L. Cohn, Sew Fun Mangano	*Researchers*
Mimi Harrison	*Illustrations Researcher*
Carolinda E. Averitt, Neil W. Averitt, Elisabeth B. Booz, Kimberly A. Kostyal	*Contributing Writers*
R. Gary Colbert	*Production Director*
Lewis R. Bassford	*Production Project Manager*
Richard S. Wain	*Production*
Meredith C. Wilcox	*Illustrations Assistant*
Gillian C. Dean, Christina Perry	*Art Production Assistants*
Peggy Candore	*Assistant to the Director*
Sandy Leonard	*Staff Assistant*
Elisabeth MacRae-Bobynskyj	*Indexer*

Manufacturing and Quality Control

George V. White	*Director*
John T. Dunn	*Associate Director*
Vincent P. Ryan	*Manager*
James J. Sorensen	*Budget Analyst*

Library of Congress Cataloging-in-Publication Data: page 271

Author John Langone, a veteran science journalist, was a staff writer for *Discover* and *Time* magazines, a reporter and writer for United Press International, and science editor at the *Boston Herald.* He was a Kennedy Fellow in Medical Ethics at Harvard, a Fellow at the Center for Advanced Study in the Behavioral Sciences at Stanford, and a Fulbright Fellow at the University of Tokyo, where he researched the impact of science and technology on the Japanese. This is Langone's 23rd book.

Art for hardcover stamp: Britt Griswold